NORTH AMERICAN PARASITIC ZOONOSES

World Class Parasites

VOLUME 6

Volumes in the World Class Parasites book series are written for researchers, students and scholars who enjoy reading about excellent research on problems of global significance. Each volume focuses on a parasite, or group of parasites, that has a major impact on human health, or agricultural productivity, and against which we have no satisfactory defense. The volumes are intended to supplement more formal texts that cover taxonomy, life cycles, morphology, vector distribution, symptoms and treatment. They integrate vector, pathogen and host biology and celebrate the diversity of approach that comprises modern parasitological research.

Cover design by D. J. Richardson
Line drawing of child and pets by Barbara Nitchke
Line drawings of *Toxocara* egg and *Toxoplasma* oocyst by D. J. Richardson
Line drawing of flea from CDC *Pictorial keys* (1996) - Full reference on pg. 203

NORTH AMERICAN PARASITIC ZOONOSES

edited by

Dennis J. Richardson
Quinnipiac University
Hamden, Connecticut

and

Peter J. Krause
Connecticut Children's Medical Center
Hartford, Connecticut
and
University of Connecticut School of Medicine
Farmington, Connecticut

Springer Science+Business Media, LLC

Library of Congress Cataloging-in-Publication Data

North American Parasitic Zoonoses / edited by Dennis J. Richardson and Peter J. Krause.

p. cm. – (World class parasites ; v. 6)

Includes bibliographical references (p.).

ISBN 978-1-4613-5405-5 ISBN 978-1-4615-1123-6 (eBook)

DOI 10.1007/978-1-4615-1123-6

1. Zoonoses—North America. 2. Parasitic diseases—North America. I. Richardson, Dennis J., 1963- II. Krause, Peter J., 1945- III. Series

RC113.5 .N67 2002
616.9'59'0097—dc21

Originally published by Kluwer Academic Publishers in 2003
Softcover reprint of the hardcover 1st edition 2003

Printed on acid-free paper.

TABLE OF CONTENTS

Preface..vii

Stealth Parasites: The Under Appreciated Burden of Parasitic Zoonoses in North America
J. L. Gauthier, Anuj Gupta, and Peter Hotez 1

Toxocariasis and Baylisascariasis
Zandra Hollaway Duprey and Peter M. Schantz..............................23

Trichinellosis
Dickson D. Despommier..41

Larval Tapeworm Infections: Cysticercosis, Cystic Echinococcosis and Alveolar Echinococcosis
Ana Flisser...57

Intestinal Tapeworm Infections
Dennis J. Richardson..73

Other Noteworthy Zoonotic Helminths
Dennis J. Richardson..85

Cryptosporidiosis
Cynthia L. Chappell and Pablo C. Okhuysen...............................113

Toxoplasmosis
David S. Lindsay, Louis M. Weiss, and Yasuhiro Suzuki..................129

Babesiosis
Paul Lantos and Peter J. Krause...151

Other Noteworthy Zoonotic Protozoa
Oscar J. Pung..165

Zoonotic Arthropod Parasites
Lance A. Durden...185

Index...205

PREFACE

Domestic animals and wildlife are important sources of food, clothing, companionship, and aesthetic enrichment but they also carry an abundance of parasites that may cause human disease. Zoonoses are a group of human diseases caused by diverse organisms that reside in animals and are transmitted to humans. Most definitions of zoonotic parasites are restricted to those utilizing a vertebrate reservoir so that diseases such as malaria are excluded. This volume addresses parasitic (helminth, protozoan, and arthropod) zoonotic diseases that affect residents of North America. Although the emphasis is on North American infections, all of the parasites discussed have a much broader geographic distribution. Therefore, this volume should prove useful to clinicians, researchers, and academicians worldwide.

Parasites serve as a source of wonder, amazement, and even beauty to the parasitologist, but one cannot forget that these fascinating creatures are also the instruments of untold human suffering. Current estimates of the world-wide parasitic health burden include 4.5 billion people infected with worms and over one million deaths each year as a direct result of worm infection. Millions more are infected and die from protozoan infections. For malaria alone, nearly a half a billion people are infected with one to two million deaths each year. The importance of parasitism is not a new revelation. The 19th century Oxford biologist George Romanes stated, "We find that more than half of the species which have survived the ceaseless struggle are parasitic in their habits, lower and insentient forms of life feasting on higher and sentient forms; we find teeth and talons whetted for slaughter, hooks and suckers, moulded for torment--everywhere a reign of terror, hunger, and sickness, with oozing blood and quivering limbs, with gasping breath and eyes of innocence that dimly close in deaths of brutal torture!"

Parasitic zoonoses are common and are not exclusively found in tropical or subtropical regions. Foodborne or waterborne zoonoses such as trichinosis and giardiasis are well recognized in North America. It is estimated that nearly 80 million people in the United States are infected by *Toxoplasma gondii*, a protozoan parasite that normally lives in the intestinal tract of cats. Although toxoplasma infections usually are asymptomatic, infection can be fatal in the fetus and other immunocompromised hosts. Other less notorious parasites exist, some of which have insidious means of infecting humans. For example, it is estimated that the common rodent tapeworm, *Hymenolepis diminuta*, infects 75 million people worldwide. A 1987 study by the Centers for Disease Control revealed that human prevalence of this tapeworm in New England varied between 0.5 and 1.6 percent. A more recent survey found that 75 percent of pet stores in southern Connecticut are selling pet rodents infected with this tapeworm.

The myriad parasitic worms and protozoans represent a large part of the overall biodiversity on Earth. The phrase, "this wormy world," coined by the parasitologist Norman R. Stoll, accurately describes both the animal and human condition.

Despite the importance of parasitic infections, medical and veterinary professionals often are poorly informed about their clinical presentation, diagnosis and management. In a recent survey of 171 physicians in Connecticut, most reported rarely encountering helminth infections and rarely discussing zoonoses and their prevention with their clients. When asked to identify two zoonotic parasites that they felt were important, 32 percent responded "don't know" and 12 percent responded "none." One of the top five zoonotic parasites listed as important were pin worms, a non-zoonotic infection! Such lack of knowledge is not surprising since fewer than half of North American medical schools require a specific course in parasitology. On average North American medical students receive just over 12 hours of instruction in parasitology.

A similar number of veterinarians were interviewed for the same survey. Although they diagnose zoonotic parasites on a daily basis in their non-human clients and were more knowledgeable than the physicians regarding zoonotic infection, most indicated that they only occasionally discussed the threat of zoonotic transmission to pet owners. When asked who should be responsible for educating the public about zoonotic diseases, both physicians and veterinarians listed public health officials as most responsible and their own profession as least responsible. The survey also revealed that communication between physicians and veterinarians regarding zoonoses is virtually non-existent. Improvement in the education of health care professionals regarding parasitic disease is of critical importance in the control of these infections.

The objective of this text is to provide a concise and useful review of essential information about parasitic zoonotic diseases for physicians and veterinarians. We hope that it will be informative for teachers, students, researchers and the general public as well. All of the authors are recognized authorities in zoonotic parasitic disease and each has adhered to the following outline in order to provide a practical and informative guide for the reader:

i. Etiology/ Natural History: Account of taxonomy, morphology as necessary for identification, and life cycles.
ii. Epidemiology: Global prevalence and prevalence in North America among various demographic groups in both human and non-human hosts. Consideration of primary risk factors associated with transmission.
iii. Pathogenesis

iv. Diagnostic tests

v. Treatment and control measures: Current drug(s) of choice, contraindications, and procedures utilized in the treatment of human infection. Consideration of practical steps to be taken in the prevention of transmission

vi. Recent Advances and Contemporary Challenges: A discussion of the most exciting contemporary areas of research concerning each disease with recommendations for future avenues of study

We are grateful to all of the authors for their illuminating work. We thank Dr. Richard Seed and Dr. Samuel Black for inviting us to serve as co-editors of this text. Finally, we are indebted to Kristen Richardson for her assistance in the preparation of this volume.

Dennis J. Richardson, Peter J. Krause, 15th June, 2002

STEALTH PARASITES: THE UNDER APPRECIATED BURDEN OF PARASITIC ZOONOSES IN NORTH AMERICA

J. L. Gauthier[1], Anuj Gupta[2], and Peter Hotez[2]

[1]Department of Epidemiology and Public Health, Yale University School of Medicine, New Haven, Connecticut
[2]Department of Microbiology and Tropical Medicine, The George Washington University Medical Center, Washington, DC

It has been estimated that 80 percent of all described human infections are caused by agents which are shared by or naturally transmitted between other vertebrate animals and humans (Schwabe, 1984). These naturally transmissible infections are called zoonoses. Many of the most important human diseases worldwide are zoonoses, as are most of the emerging infections recognized in the past half-century. Of nine infectious epidemic diseases that WHO considers "either new or volatile or pose an important public health threat", five have known animal reservoirs (WHO, 2000). A recent analysis ascribes 73 percent of human emerging infectious diseases to zoonotic pathogens (Taylor and Woolhouse, 2000).

It seems that North Americans view themselves spared and somehow isolated from "this wormy world", a phrase coined by parasitologist Norman Stoll in 1947 (Stoll, 1947). However, the notion that human parasitic disease is weird or foreign to the developed world is an illusion. There are many "native-born" parasitic infections in the United States, including an estimated 55 million worm infections (Roberts and Janovy, 1996). The CDC estimates that nearly one out of four American adults has been exposed to *Toxoplasma gondii* (CDC, 2000a). At least 400,000 residents of Milwaukee, Wisconsin were parasitized in a single massive outbreak of cryptosporidiosis in 1993 (MacKenzie et al., 1994).

Several reasons for this under-appreciation of the public health impact of parasites have been cited: many parasitic diseases are underdiagnosed; these infections disproportionately affect the poor and therefore receive little attention; and even that parasitism is not an appropriate topic for social conversation (Roberts and Janovy, 1996). Parasitic infections endemic to North America cause significant human morbidity and mortality, though there is evidence that at least three of these infections are grossly under-recognized as public health problems. Despite the availability of diagnostic screening tools, toxocariasis, neurocysticercosis, and toxoplasmosis continue to cause severe, long-term disability, often beginning in childhood.

International movement of people and animals quickly distributes parasites throughout the globe, transporting unusual infections into North America. Under the right conditions, such as climate and the presence of vectors, these "foreign-born" infections could become locally established in humans, domestic animals, and wildlife. The number of people living with immunodeficiency continues to rise, increasing the proportion of the population especially susceptible to zoonotic parasitism.

In 1969, the U.S. Surgeon General, William H Stewart, enthused by success against typhoid, cholera and smallpox, told Congress that it was time to "close the book on infectious diseases, declare the war against pestilence won, and shift national resources to such chronic problems as cancer and heart disease." (WHO, 2000) This now infamous quote foreshadowed an era of complacency in communicable disease surveillance and control. In the 30 years since Surgeon General Stewart's message to congress, more than 30 new infectious diseases have been recognized and a resurgence of old threats such as malaria and cholera has occurred. Other pathogens have appeared in new places, exemplified by the arrival of West Nile virus in the northeastern United States in 1999.

Emerging diseases have been defined as "new, re-emerging, or drug-resistant infections whose incidence in humans has increased within the past two decades or whose incidence threatens to increase in the near future" (IMCEMTH, 1992). Parasitic zoonoses are found among the emerging infections identified since 1970. The list includes the first case of human babesiosis in the United States reported in 1968 (Scholtens et al., 1968); *Cryptosporidium parvum*, identified in 1976; and *Baylisascaris* eosinophilic meningoencephalitis, documented in 1984.

For these reasons, it is necessary that physicians have some familiarity with parasitic disease; more and more frequently the practitioner is faced with these "exotic" infections.

POPULATIONS AT RISK FOR INFECTION WITH ZOONOTIC PARASITES

Immunocompromised persons

The most severe manifestations of parasitic zoonoses are seen in immunocompromised individuals. Opportunistic infections have been described in both children and adults with immune system impairment due to pathologic processes such as human immunodeficiency virus, congenital immunodeficiency syndromes, hematopoietic malignancies, and diabetes, or because of medical treatments that induce immunosuppression, such as organ transplantation or cancer chemotherapy. This particularly susceptible proportion of the population is continually increasing, as population growth and medical advances increase the number of people living with immunosuppressive conditions. It is estimated that well over one million new cancer cases were diagnosed in 1999 alone. Projections of prevalence rates yield an estimated 6.2 million people with a history of cancer in the United States by the year 2000 and 9.6 million by 2030. Due to highly active antiviral therapy delaying the onset of AIDS in HIV positive individuals, new cases reported in the United States have decreased. The number of people living with AIDS, however, is increasing as new treatments prolong the lives of infected persons. More than 300,000 people were living with AIDS in 2000, an 18% increase from 1998 (CDC, 2002). As of October 1, 2001, there were 78,350 people on the U.S. national organ transplant waiting list, approximately three times the number of transplants (23,360) performed in that year (UNOS, 2001).

More than 100 pathogens, ranging from viruses to arthropods, have been associated with opportunistic infections in HIV-infected persons; a few of these diseases are parasitic zoonoses that cause high rates of morbidity and are considered AIDS-defining illnesses (CDC, 1997). The incidence of *Cryptosporidium parvum* in AIDS patients has been reported to be as high as 5 to 10% per year. Cerebral toxoplasmosis is one of the most common AIDS indicator conditions reported in the United States. Chagas' disease, caused by *Trypanosoma cruzi*, and visceral leishmaniasis are both exacerbated by the immunosuppression of HIV/AIDS. Leishmaniasis is becoming an important opportunistic infection even in areas where the disease is not normally endemic.

In transplant recipients, latent infection of transplanted tissues with *Toxoplasma gondii* may be a significant problem, resulting in severe, overt toxoplasmosis. Twelve to 17 percent of patients in a British study became infected as a result of their transplant, and among these patients, half developed clinical toxoplasmosis (Wreghitt et al., 1986). On the other hand, toxoplasmosis associated with bone marrow transplantation appears to be due largely to reactivation of latent infection in the recipient.

The only intestinal nematode that is known to be an opportunistic infection and to present special problems in immunocompromised hosts is *Strongyloides stercoralis*. Disseminated strongyloidiasis, or "hyperinfection syndrome", has been reported among patients receiving high dose corticosteroid therapy (Liu and Weller, 1993) and among HTLV-1 infected patients.

Children

For some good reasons, contact with domestic animals is viewed as an enriching feature of childhood. Not only do pets provide important sources of social support and emotional attachment for many children, but domestic animals have been shown to help develop nurturing behavior and nonverbal communication skills (McNicholas and Collis, 2001). Most children in North America have regular contact with companion animals; between 60 to 80%

of families with children aged 6 –14 keep at least one pet in the household (Melson et al., 1997).

Besides the fact that children are commonly and frequently exposed to animals, several other childhood behaviors increase a child's risk of zoonotic infection. Behaviors normally seen at various stages of development, such as mouthing objects, playing in dirt, and a relative lack of hygiene, expose children to soil- and fecally-transmitted pathogens. Pica, the compulsive eating of non-food substances, and geophagia, the most common form of pica, are frequently observed in children, and are known risk factors for toxocariasis and other nematode parasitisms (Glickman et al., 1981). Children with mental disabilities may exhibit pica and geophagia more frequently, and to a greater extent than children without impairments. A large serosurvey of mentally retarded children revealed high seroprevalence rates for *Toxocara canis*, *Strongyloides stercoralis*, and *Entamoeba histolytica* (Brook et al., 1981). One institutionalized child with a history of pica was diagnosed with three concurrent parasitic infections: visceral larva migrans, *Entamoeba coli*, and latent toxoplasmosis (Marcus and Stambler, 1979).

Occupational and Recreational Exposure

Not surprisingly, occupations involving close and frequent contact with animals have high rates of exposure to zoonotic parasitic disease. Animal caretakers, farmers, slaughterhouse workers, and veterinarians should be considered personnel at increased risk for acquiring infections. A Wisconsin survey showed that dairy farmers are nearly twice as likely to be seropositive for *Cryptosporidium parvum* than other persons (Lengerich et al., 1993). Over six percent of 152 Austrian swine farmers had serum antibodies against *Echinococcus* antigens; in comparison, 50 control subjects with no exposure to farm animals and 137 veterinarians were all sero-negative (Deutz et al., 2000). British farmers and Austrian veterinarians had higher rates of exposure to *Toxoplasma gondii* than the general population (Thomas et al., 1999; Nowotny and Deutz, 2000).

Serosurveys for *Toxocara canis* exposure have not consistently demonstrated occupational associations, probably as a result of regional variations in transmission and human behavior. For example, studies of animal caretakers (Slovak, 1984) and veterinarians (Glickman and Cypress, 1977) in Europe and North America showed no increased risk of infection compared to the general population, but dog breeders in Britain did have significantly elevated antibody levels (Woodruff et al., 1978). A recent Austrian survey revealed that toxocaral antibody prevalence was 20 times higher among veterinarians than in the general Austrian population (Nowotny and Deutz, 2000).

Outdoor recreational activities may carry some risk of exposure to parasitic zoonoses. Although campers and backpackers are described as suffering from a high incidence of *Giardia lamblia* diarrhea, a meta-analysis performed by the CDC failed to find an association with drinking from wilderness water sources (Welch, 2000). In addition, confirmation of animal sources of oocysts in connection with campers' giardiasis has not been made. An outbreak of cryptosporidiosis occurred in persons using a recreational lake at a state park in New Jersey; again, the source of infective oocysts was not found (Kramer et al., 1998). Protozoal diarrhea outbreaks associated with swimming pools likely involve human-to-human, rather than zoonotic transmission. In a few instances, the animal source for zoonotic transmission during recreational activities has been clearly demonstrated. For example, in 1977, an outbreak of toxoplasmosis occurred among patrons of a riding stable, where infected cats were implicated as the source of infection (Teutsch et al., 1979).

Swimmer's itch, caused by bird schistosomes, and cutaneous larva migrans, caused by canine hookworms, are miserable souvenirs of a lake or seashore visit. In the United States, swimmer's itch mainly plagues tourists to the Great Lakes area, but it has been reported from nearly all states (Roberts and Janovy, 1996). Cutaneous larva migrans, or creeping eruption, is typically a tropical problem, but occasional cases have been acquired on beaches as far north as Long Island, New York (Esser et al., 1999).

Immigrants and Travelers

In the past 50 years, international travel has increased dramatically. The number of international tourist arrivals worldwide is expected to increase from 5 million in 1950 to 937 million in 2010 (Paci, 1995). The WHO estimates that 1.4 billion people travel by air every year (WHO, 1999). Within a matter of hours, infectious agents or their vectors can travel to virtually any part of the globe. Traveler's diarrhea is the most important health problem of tourists; the incidence is nearly 40 percent in visitors to developing nations (Ericsson, 1998). A significant percentage suffer from gastointestinal illness caused by the parasitic zoonoses *Giardia lamblia*, *Cryptosporidium parvum*, and *Entamoeba histolytica* (Okhuysen, 2001). A wide variety of zoonotic helminthiases, tick-borne infections, and myiases have been documented in travelers returning from abroad. Recently, a large outbreak of eosinophilic meningitis caused by the rat lungworm, *Angiostrongylus cantonensis,* was reported; 12 U.S. travelers returning from Jamaica were affected (Slom et al., 2002).

Foreign animal imports, including pets, zoo and laboratory animals, and illegal trade, amount to more than U.S. $1 billion annually (Barton, 1996). Parasite pathogens or their vectors may accompany animals moving internationally. In 2000, a horse imported from Argentina to Florida was found to be infested with screwworm larvae (AVMA, 2000). Screwworm (*Cochliomyia hominivorax*), the most important cause of myiasis in the world, was eradicated from the United States in 1966 at considerable expense. The USDA estimates that livestock production losses could approach $750 million annually if the parasite were to become re-established (Rogers and Redding, 2000). Screwworms in humans are not uncommon, causing dermal and nasopharyngeal myiasis.

Between 500,000 and 1,500,000 legal aliens entered the United States annually during the 1990's (INS, 1995). Cultural practices in certain immigrant communities, particularly consumption of raw or undercooked meat, have been associated with outbreaks of zoonotic disease. A traditional pork dish at a Laotian wedding in Iowa was responsible for 90 cases of trichinosis (Stehr-Green and Schantz, 1986). Indigenous transmission of the pork tapeworm, *Taenia solium*, appears to have occurred among immigrants living in California (Ehnert et al., 1992.).

Parasitic infections can persist in immigrants and in travelers returning from abroad long after their arrival in North America. For instance, the Minnesota Department of Health screened 71 East African refugees for enteric parasites; roughly 25 percent of people who submitted fecal specimens carried pathogenic parasites, and most of these infected individuals had resided in the U.S. for greater than one year (Sachs et al., 2000). Cases of 40-year old *Stongyloides stercoralis* infections have been documented in ex-prisoners of war (Roberts and Janovy, 1996).

International adoption, representing an intersection of two at-risk groups, is an increasingly common phenomenon in the United States and Canada. Immigrant visas issued to orphans entering the United States (a measure for international adoptions) totaled 13,620 in 1997, more than double the amount in 1992 (Lears et al., 1998). Adoptees frequently arrive carrying parasitic infections endemic in the childrens' country of origin, creating unusual challenges for primary care physicians. In children adopted from Romania, intestinal parasites were found in 33 percent of patients, and 45 percent of infected children had two or more pathogens (Johnson et al., 1992). Nine percent of Chinese adoptees were diagnosed with enteric parasite infections, of which *Giardia* was the most frequent identified pathogen (Miller and Hendrie, 2000). In children immigrating to North America, a thorough medical history is imperative in assessing the risk of parasitic disease. The history must include questions pertaining to the child's residential history and living circumstances; the circumstances of the child's immigration; the quality and quantity of the child's diet and medical care; and the child's medical history.

FACTORS INFLUENCING TRANSMISSION OF PARASITIC ZOONOSES IN THE DEVELOPED WORLD

Human behavior is probably the most important factor determining an individual's exposure to parasitic zoonoses. Parasites are not randomly distributed throughout the globe. Human ecology, agricultural practices, and cultural beliefs are extremely influential on the rate and pattern of parasitic infections in communities and regions. Hunting, farming, keeping pets, and a variety of other occupational and recreational contacts with other vertebrates exposes people to the parasites of an enormous variety of animal species. Our alterations to the environment have consequential effects, sometimes predicted but often unforeseen, on host and parasite populations.

Companion Animals

The number of companion animals kept by U.S. families continues to increase every year. Roughly 58 percent of families reported one or more pets in their households in 1996 (AVMA, 1997), and most people will own a pet at some time during their lives (Gammonley, 1991). Pet ownership is quite common among people considered most at risk for acquiring zoonotic disease, including children and immunocompromised people. Households most likely to have a pet are families with children aged 6-14 (AVMA, 1997). A 1993 survey of 408 adults with AIDS revealed that 46 percent had lived with a companion animal during the previous five years (Conti et al., 1995).

Fortunately, pet-transmitted infections are rare in the developed world today. Except for *Bartonella henselae* (the cause of cat scratch fever) and zoophilic dermatophytes (ringworm), zoonoses are more commonly acquired from sources other than animals kept as pets. Pets have been estimated to account for only 3 percent of Salmonella outbreaks (CDC, 1977). Less than 7 percent of *Campylobacter jejuni/coli* cases are attributed to animal contact (Saeed et al., 1993). In a large study (over 700 patients), Wallace et al. (1993) found no increase in *Toxoplasma gondii* infection rates among HIV-infected cat-owners over HIV-infected persons who did not keep cats. Among HIV-positive cat owners, seroprevalence to *Toxoplasma* was equivalent to overall rates of exposure in other populations nationwide. Glaser et al (1998) investigated the association between *Cryptosporidium* infection and animal exposure in HIV-seropositive individuals. In the case-control study, no significant difference in the rate of pet ownership was found between persons with and without cryptosporidiosis.

Some clinicians advocate removal of all domestic animals from households with family members at special risk for acquiring zoonotic disease, especialy immunocompromised individuals and pregnant women (Spencer, 1992). Some patients give up their pets out of fear of disease transmission (Burton, 1989). However, relinquishing companion animals based on concern over zoonotic disease transmission may negatively impact patients' overall health status. Pets provide their owners with psychological, social, and physical benefits; the positive effect of the human-animal bond on human health has been demonstrated repeatedly (Friedmann et al., 1980; Anderson, 1992; Reade,1995).

Food Animals

North American agriculture changed dramatically during the 20th century, with resources becoming concentrated into fewer and larger farms. While U.S. farm production has doubled over the last fifty years, farm numbers dropped by more than two-thirds. Large commercial farms account for just 8 percent of all farms, but produce 68 percent of total agricultural output (USDA, 2001). The shift towards more intensive farming practices is linked to emerging public health and food safety concerns, including parasitic zoonoses.

Confined animal operations offer the opportunity to control some zoonotic pathogens that have historically plagued agricultural societies, though the formidable task of animal waste handling from these facilities creates other public health problems. At these farms, a single building may house thousands of animals from birth to slaughter, eliminating any exposure to the outdoor environment. *Trichinella*-free pig farming is a feasible option for controlling this zoonosis, even in endemic areas such as North America (van Knapen, 2000). In confined swine operations, which raise a greater percentage of North American hogs every year, pigs are denied access to carrion, garbage, and carcasses, limiting exposure to infection with *Trichinella.* Since 1947, the annual incidence of reported human trichinosis cases has declined from 300-400 cases a year to 100-150 (CDC, 1982). Total confinement systems have been shown to virtually eliminate both *Toxoplasma gondii* and *Trichinella spiralis* infection from pigs (Davies et al., 1998).

Cryptosporidiosis has emerged as an important cause of diarrhea in North America, with serious, potentially fatal infections occurring among immunocompromised individuals. Giardiasis, the most frequent culprit of waterborne outbeaks of infectious diarrhea, affects an estimated 2.5 million U.S. citizens annually (Olson, 2002). Because both parasites infect a variety of domestic animals and are prevalent in livestock, agricultural sources are thought to play a role in surface water contamination. However, there is little research to document the true sources of oocysts in drinking water supplies or to identify agriculture as the primary source (Sischo et al., 2000).

In adult cattle, *Cryptosporidium parvum* is a common, usually asympotmatic infection; in calves, the parasite is highly prevalent and may cause severe disease. A nationwide survey in the United States revealed that 59 percent of cattle farms had *C. parvum*-infected animals (Garber et al., 1994). The prevalence of *C. parvum* infection in calves exceeded 80 percent in a Canadian study, and rates as high as 92 percent were demonstrated in California calves (Olson et al., 1997; Atwill et al., 1998). Studies have shown increased oocyst concentrations downstream of cattle farms, and certain agricultural practices (e.g. cows having free access to streams, frequent manure spreading on fields) have been identified as risk factors for watershed contamination (Sischo et al., 2000).

The prevalence of *Giardia lamblia* infection in food animals in the U.S and Canada ranges from 23 to 89 percent in calves, 34 to 82 percent in lambs, and 7 to 10 percent in pigs (Olson, 2002). Recently, molecular techniques have subdivided *G. lamblia* isolates into genotypes with varying host specificity. Certain genotypes readily infect both humans and livestock, adding justification to the concern that manure may contaminate surface waters and drain into municipal water reservoirs. One particular waterborne giardiasis outbreak in humans has been attributed to pasture run-off leading to drinking water contamination (Furness et al., 2000). The risk of *Giardia* infection is known to increase with high population densities (Olson, 2002), as seen in modern confined animal operations. In addition, intensive farming practices (e.g. feedlots, large confined swine and poultry facilities) can have a major impact on local water quality due to high livestock densities generating large volumes of animal waste.

Wildlife

Wildlife populations have been considered a "zoonotic pool", forming reservoirs from which zoonoses may emerge to threaten public health. The impact of human activity on wildlife populations is clear; unprecendented rates of wildlife habitat destruction, increasing human encroachment into wildlife territories, fragmented habitats, and loss of biodiversity are well documented. However, the link between anthropogenic changes to wildlife habitat and human disease emergence is not well described.

A number of underlying factors are thought to increase the risk of "spillover" of zoonotic disease from wildlife reservoirs to humans. International movement of animals, both wild and domestic, introduces animal pathogens to new environments. Human and domestic animal population expansion reduces available wildlife habitat and increases contact between wildlife and people. Ill-defined environmental "stressors" created by human activity are hypothesized to cause immunosuppression and increased disease susceptibility in wild populations (Daszak et al., 2001). Isolation and concentration of wild species through game reserves, blocked natural migration routes, and habitat fragmentation has been shown to increase disease transmission rates (Dobson and May, 1986). Some wild animal populations are thriving from their close association with humans. A recent field investigation of a case of human baylisascariasis in California revealed an extreme population density of 30 raccoons per one-quarter acre (Park et al., 2000). Absence of predators and epizootic disease outbreaks (such as canine distemper and rabies), and the widespread availability of pet food left outdoors are factors contributing to the increase in the number of suburban raccoons.

Environmental Alterations

Human modifications to the environment have brought about major increases in parasitic disease. Construction of dams, with resultant flooding, has created new habitats for the snail host of schistosomiasis and caused an upsurge of the disease. After the completion of Egypt's Aswan dam, a tremendous increase in schistosomiasis was seen in people living around Lake Nasser. Explosive outbreaks of schistosomiasis have also been seen following the construction of major water-control projects in Ghana and Senegal (Aksoy et al., 2000).

It is evident that the impact of these large-scale projects on the ecology of a region should be assessed prior to construction, rather than after the damage is done. McSweegan (1996) has proposed the use of an Infectious Disease Impact Statement to predict local changes in infectious diseases caused by human-engineered activities. A risk assessment of Turkey's GAP irrigation project by a multi-national group of scientists (Aksoy et al., 2000) anticipated increased rates of leishmaniasis, malaria, amebiasis and giardiasis, and warned about the emergence of schistosomiasis in the 75,000 sq. km affected area.

One of the many hypothesized and vigorously debated effects of global warming is the eventual spread of tropical disease to temperate climates. Killick-Kendrick (1996) suggests that visceral leishmaniasis may become endemic in southern England, based on a prediction that by 2025, the area will have a Mediterranean climate capable of supporting the sandfly vector. Several mathematical models predict that mosquito-borne diseases, particularly malaria, will increase in northern climates if the warming trend continues; other authors reject these simulations (Rogers and Randolph, 2000). Two locally-acquired cases of malaria occurred in Suffolk County, New York in 1999 during one of the warmest and driest summers ever recorded in the northeastern United States (CDC, 2000b).

TRENDS IN MEDICAL PARASITOLOGY EDUCATION IN THE UNITED STATES

Physicians, osteopaths, physician assistants, nurse practitioners, and other health care professionals in North America are largely unaware of the public health threat from zoonotically transmitted helminths and protozoa. An important reason for this situation is the absence of qualified trained parasitologists in American schools of medicine, osteopathic medicine, and nursing. Students in the health professions receive remarkably little didactic instruction on the topic of parasitic disease and typically no laboratory instruction; in some North American professional schools, parasitology is not taught at all. Many curriculum coordinators erroneously and naively assume that parasitic infections are restricted to the developing nations of the tropics and therefore not relevant to health professionals who are unlikely to practice outside of North America. Given the lofty standard of living, strong economies, and superior methods of waste disposal found in North America, many in the medical and public health fields are not aware of the emergence of helminthic and protozoan infections in their own communities. Our globalized economy is slowly contributing to a revamping of this attitude, but substantive changes in American medical and public health education have not yet taken place.

Observations on medical parasitology education might explain some of the alarming trends about the knowledge base of health care practitioners in the United States. For instance, fewer than 30 percent of pediatricians in Connecticut are aware that the diagnosis of cryptosporidiosis requires that the health care provider request a specific diagnostic test to identify the parasite oocysts (Morin et al, 1997).

In contrast to American medical education, colleges of veterinary medicine do provide significant numbers of teacher-student didactic hours in parasitology. Virtually every North American Veterinary Medical College retains one or more parasitologists on their staff. Unfortunately, there are few if any mechanisms by which North American-trained veterinarians have the opportunity to convey their knowledge to physicians, osteopaths, PA's, and nurses. With some exceptions, most American schools of allopathic and osteopathic medicine are not in geographic proximity with veterinary medical colleges; the former are often located in large urban medical centers while the latter are placed in more rural settings on the campuses of large land-grant universities. There are some interesting exceptions to this paradigm (Table 1) which might eventually afford opportunities for improved communication between animal and human health professionals, perhaps leading to innovative programs in emerging zoonoses or joint MD-DVM, DO-DVM, PA-DVM, and MSN-DVM programs.

Table 1. New opportunities for zoonoses education: schools where North American medical and veterinary education overlaps.

Michigan State University	East Lansing, MI
Ohio State University	Columbus, OH
Texas A&M University	College Station, TX
Tufts University	Boston, MA
University of California	Davis, CA
University of Florida	Gainsville, FL
University of Illinois	Urbana-Champaign, IL
University of Minnesota	Minneapolis, MN
University of Pennsylvania	Philadelphia, PA
University of Saskatchewan	Saskatoon, SA
University of Wisconsin	Madison, WI

NEED FOR NATIONAL SURVEILLANCE OF PARASITES

Our public health infrastructure is partly to blame for the general absence of awareness of parasitic zoonoses in North America. Several pathogens remain underappreciated as causes of disease: *Cryptosporidum*, which accounts for large water-borne outbreaks of gastroenteritis in the U.S.; *Toxocara*, which may now arguably represent the most prevalent helminthic infection in the U.S.; and *Taenia solium*, which causes cysticercosis, a leading cause of epilepsy in Mexico and the Southwestern United States. Every year, infants develop postnatal blindness and mental retardation because they are not screened in the newborn period for toxoplasmosis; only the state of Massachusetts conducts newborn screening for vertically acquired *Toxoplasma* infection.

Current surveillance data that might give us the basis for implementing appropriate public health practice and control is strongly lacking. For instance, a recent study conducted in Bridgeport, Connecticut, revealed that at least 25 percent of Hispanic children ages 2-15 were seropositive for infection with *Toxocara canis*, a zoonotic ascarid found in the feces of dogs (Sharghi et al, 2001). *Toxocara* is responsible for producing visceral larva migrans and ocular larva migrans in humans and has recently been implicated as a possible causative factor in childhood asthma (Buijs et al, 1994). Could *Toxocara* partly account for the latest increase in pediatric asthma among inner-city Hispanic children? Until this time, infection with *Toxocara* has been considered second only to pinworm as the most prevalent helminth infection in North America. It is now suspected, however, that *Toxocara* may have surpassed pinworm in prevalence. Without further detailed studies, it is not possible to ascertain the validity of either suspicion.

Similarly, cysticercosis is now believed to be a leading cause of seizures in North American children. The disease is being increasingly documented in US cities with high populations of Mexican and Central American immigrants, such as Los Angeles and San Antonio. The public health implications of cysticercosis become apparent when statistics are taken into consideration. Greater than 90 percent of people in the United States with cysticercosis are foreign-born (Hotez, 2002). In the state of California, immigrants constitute 26 percent of the population, predominantly from Mexico and Latin America. Furthermore, nearly 200,000 children entered the United States in 1998, with an additional 30,000 entering illegally, many of whom came from these nations. Given the increasing number of immigrants from nations in which *Taenia solium* is endemic, it is imperative that public health practitioners become more educated about this illness. At this time, however, despite the overwhelming implications of this disease, we have very little information concerning its actual prevalence.

Implementing a North American survey comprised of fecal examinations and newborn screening would likely reveal, for the first time, the enormous economic and health impact of human parasites. It is becoming increasingly important for our societies to expand upon the growing interest in parasitic disease in North America by educating both the public and health practitioners about indigenous and imported human parasites.

Table2. Possible ranking of the most common parasitic infections in the United States and Canada

Helminth Infections	*Protozoan Infections*
1. Toxocariasis*	1. Giardiasis*
2. Enterobiasis	2. Toxoplasmosis*
3. Taeniasis (Neurocysticercosis)*	3. Dientamoebiasis
	4. Cryptosporidiosis*
	5. Blastocystosis

* zoonotic transmission reported
(Modified from Hotez, 2002)

THE ROLE OF VETERINARY MEDICINE IN PREVENTING ZOONOTIC DISEASE

Zoonotic disease control requires involvement of both physicians and veterinarians. Human medicine does not deal with animal reservoirs or their role in transmission of zoonoses, and veterinary medicine is not concerned with the diagnosis and treatment of human disease. Because each profession can affect some, but not all, of the links in the chain of infection, communication between physicians and veterinarians facilitates prevention of human zoonotic disease.

Unfortunately, surveys have shown that communication between physicians and veterinarians is nearly non-existent. In a recent survey of Connecticut practitioners, both veterinarians and pediatricians indicated that they rarely or had never consulted their counterparts regarding zoonoses. Sixty percent of pediatricians had never asked a veterinarian for information about a zoonotic disease, and conversely, 41 percent of veterinarians had never sought advice about the topic from a physician. In addition, pediatricians rarely contacted other physicians or academicians for advice regarding zoonoses, though they did call upon other professionals more frequently than they contacted veterinarians (Gauthier and Richardson, 2002). The lack of communication between the professions was also demonstrated in a survey of Wisconsin veterinarians and physicians (Grant and Olsen, 1999).

To further complicate the disconnect between professionals, there is disagreement between physicians and veterinarians about which profession has primary responsibility in advising clients about prevention of zoonoses. In the Connecticut practitioner survey, the two groups of professionals were asked to rank the relative importance of four occupations (animal control officer, physician, public health official, or veterinarian) with respect to their responsibility for informing the general public about zoonotic disease prevention (Gauthier and Richardson, 2002). On average, veterinarians thought that physicians had primary responsibility for public education, and that public health officials played the second most important role. Physicians most frequently placed highest responsibility with public health officials, and they felt that veterinarians were next most important. Grant and Olsen (1999) also found that physicians expected veterinarians to play an equal or greater role in advising patients about zoonotic disease.

Overall, veterinarians feel more comfortable than physicians in advising clients about the risks of zoonotic disease. While 45 percent of veterinarians responded that they felt very comfortable discussing zoonosis prevention with their clients, only 6 percent of pediatricians felt as confident in this advisory role (Gauthier and Richardson, 2002). Veterinary practitioners may feel more confident talking about animal-transmitted disease because they encounter zoonoses in their patient population much more frequently than primary care physicians. Typically, veterinarians diagnose zoonoses in animals on a weekly basis, while physicians report that they encounter human zoonotic disease only occasionally (Grant and Olsen, 1999).

Veterinarians have always played an important role in safeguarding human health and the human-animal bond (Schwabe, 1984); however, recent surveys show that companion animal practitioners could be more proactive in protecting their clients' well-being. Despite veterinarians' confidence in discussing animal-transmitted infections, many pet owners do not recognize the veterinary profession as a primary source of advice about the risks of zoonotic infection. Although roughly one-third of surveyed veterinarians offer special consultation for clients concerned about pet-associated infections, only 21 percent of HIV patients felt most comfortable in asking their veterinarian about the health risks of pet ownership (St. Pierre et al., 1996). More than two-thirds of surveyed veterinarians never talk to clients about their health status (Grant and Olsen, 1999). In addition, only one in five small animal practitioners follows CDC recommendations for preventing transmission of zoonotic helminths from pets to people (CDC, 1995). Companion animal practitioners can help prevent human zoonotic disease in several ways: by providing optimal preventive medicine; by communicating with clients, physicians, and the public health community; and by educating themselves on current zoonosis issues.

Animals immunodepressed by preventable diseases are more likely to harbor zoonotic co-infections, and may possibly shed large numbers of infective organisms over a protracted period of time. Routine examinations with vaccination, prophylactic deworming, and ectoparasite control make pets safer family members. CDC guidelines for the strategic deworming of dogs and cats should be incorporated into practice protocols (CDC, 1995). Similarly, food animal veterinarians enhance herd immunity by advising farm managers on strategic vaccination and deworming programs, nutrition, stocking densities, ventilation, and other husbandry issues.

It should be emphasized that pet-transmitted infections are rare, and in general, are easily preventable. Clients can protect themselves from zoonotic infections through common-sense precautions. Handwashing, cooking meat thoroughly and handling food properly, and preventing animals from defecating in children's play areas are important, basic measures. Children should be taught to wash their hands frequently and to handle pets gently. Dogs and cats should not be allowed to roam, hunt, or eat carrion.

Part of the dialogue during well-animal visits should include a discussion of special circumstances in the family that may make zoonosis transmission more likely, particularly if the pet is living with young children or immunocompromised individuals. Client information brochures about pet-transmitted infections are available from the CDC; these can be placed in conspicuous areas such as exam rooms and waiting areas (Table 3).

Cat owners frequently ask their veterinarians about the risk of individual pet cats for the zoonotic transfer of *Toxoplasma gondii*. The risk of acquiring toxoplasmosis from a pet cat is generally overstated by medical professionals; it should be known that ingestion of undercooked meat is a much more important source of human infection than contact with cats (CDC, 2000a). Cat owners can greatly reduce their risk by using litterbox liners and cleaning the litterbox daily. If possible, immunocompromised people or pregnant women should leave litterbox cleaning duties to other family members. To prevent cats from becoming infected with *T. gondii*, cats should not be fed raw or undercooked meat or be allowed to hunt.

Veterinarians should discourage at-risk clients from purchasing and adopting certain types of pets. Immunocompromised individuals are advised not to bring home young puppies and kittens, especially sick or diarrheic animals, due to the risk of enteric infections particularly *Cryptosporidium* and *Giardia*. HIV-infected people who wish to assume the small risk for acquiring a puppy or kitten aged less than 6 months are advised to have their veterinarian examine the animal's stool for *Cryptosporidium* before they have contact with the animal (CDC, 2000c). Wild animals are not suitable pets for anyone. Raccoons living closely with people are particularly dangerous, in part because of the potential for serious disease caused by *Baylisascaris procyonis*.

Food animal veterinarians have an essential role in advising managers of intensive livestock operations on public and environment health risks. Waste management plans are intended to reduce microbial contamination of the watershed; methods of effluent treatment, storage, and handling can decrease pathogen concentrations and reduce the volume of agricultural runoff. Facilities should be designed to prevent animals from having direct contact with streams, and exclude potential intermediate hosts, such as *Trichinella*-infected rats. Farm workers themselves are at greater risk of acquiring a zoonotic infection, and veterinarians should discuss ways to avoid transmission. Drinking raw, unpasteurized milk, or eating in animal enclosures are examples of hazardous behaviors.

Any public health advice given by veterinarians should be noted on the client's record. Clients who think they have contracted a zoonotic disease should always be referred to a physician; it is outside the role of the veterinarian to diagnose or suggest treatment for human zoonotic infections.

Veterinarians can facilitate communication by offering to speak to a client's physician about zoonotic transmission concerns. Physicians may request that a potential transplant recipient's pet receive a thorough veterinary evaluation; CDC guidelines are available to help veterinarians screen companion animals for human health hazards (CDC, 2000c)(Table 3).

Despite the medical community's growing recognition of the physical and emotional benefits of the human-animal bond, some practitioners may still urge patients to give up a healthy companion animal. Veterinarians should be prepared to discuss known pet-associated risks and benefits with both pet-owner and physician. In most instances, the CDC asserts, "You do *not* have to give up your pet" (CDC, 1997).

Veterinary clinicians should not hesitate to contact local public health authorities if they witness unusual occurrence of infectious disease in pets, livestock, or wildlife. Animals frequently serve as sentinels for potential outbreaks of zoonotic disease in humans. Surveillance systems are inadequate in many regions, and therefore, the first alert that public health officials receive is often through field reports from private practitioners. For example, the identification of West Nile virus in the northeast U.S. was helped by the persistence of veterinarians in obtaining a specific etiologic diagnosis for a cluster of avian deaths (CDC, 1999).

Continuing education is important to maintaining and updating veterinarians' knowledge of local zoonotic threats. Practitioners should be aware of local infectious disease risks, such as enzootic foci of echinococcosis or the prevalence of 'creeping eruption' at local beaches. Veterinarians should also check periodically for changes to animal control regulations and reportable disease lists. Online subscriptions to ProMED Mail (www.promedmail.org) or veterinary databases can alert practitioners to the latest human and animal disease topics.

Table 3. Resources for Veterinarians

Guidelines:

- Recommendations for Veterinarians: How To Prevent Transmission of Intestinal Roundworms from Pets to People
  http://www.cdc.gov/ncidod/diseases/roundwrm/roundwrm.htm

- 1999 USPHS/IDSA Guidelines for the Prevention of Opportunistic Infections in Persons Infected with Human Immunodeficiency Virus
  http://www.cdc.gov/mmwr/preview/mmwrhtml/rr4810a1.htm

- Guidelines for Preventing Opportunistic Infections Among Hematopoietic Stem Cell Transplant Recipients
  http://www.cdc.gov/mmwr/preview/mmwrhtml/rr4910a1.htm

- Preventing Zoonotic Diseases in Immunocompromised Persons: The Role of Physicians and Veterinarians
  http://www.cdc.gov/ncidod/eid/vol5no1/grant.htm

Client Information Brochures:

- Infectious Disease Information: Diseases Related to Pets
  http://www.cdc.gov/ncidod/diseases/pets/index.htm
- Preventing Infections from Pets: A Guide for People with HIV Infection
  http://www.cdc.gov/hiv/pubs/brochure/oi_pets.htm

REFERENCES

Aksoy S, Ariturk S, Martine Y.K et al. 2000. Letters: The GAP project in southeastern Turkey: the potential for the emergence of diseases. *Emerg Inf Dis* 1(2): 62.
Anderson, WP. 1992. Pet ownership and risk factors for cardiovascular disease. *Med J Aus*157: 298-301.
Atwill ER, Harp, JA, Jones T, et al. 1998. Evaluation of periparturient diary cows and contact surfaces as a reservoir of *Cryptosporidium parvum* for calfhood infection. *Am J Vet Res* 59: 1116-1121.
AVMA. 1997. *U.S. Pet Ownership and Demographics Source Book.* Schaumburg, IL: AVMA.
AVMA. 2000. Screwworm turns heads in Florida. *JAVMA* 216(8): 1200.
Barton DR. 1996. United States Foreign Trade Highlights, 1996. Washington, D.C.: Government Printing Office.
Brook I, Fish CH, Schantz PM, Cotton DD. 1981. Toxocariasis in an institution for the mentally retarded. *Infection Control. 2(4):317-20..*
Burton BJ. 1989. Pets and PWAs: claims of health risk exaggerated. *AIDS Patient Care* 3: 34-37.
Buijs J, Borsboom G, van Gemund JJ et al. 1994. Toxocara seroprevalence in 5 year old elementary schoolchildren: relation with allergic asthma. *Am J Epi* 140: 839-47.
CDC. 1977. *Salmonella surveillance, annual summary 1976.* Washington, DC: USPHS.
CDC. 1982. Epidemiologic notes and reports common source outbreaks of trichinosis –New York City, Rhode Island. *MMWR* 31(13): 161-164.
CDC. 1995. Recommendations for veterinarians; how to prevent transmission of intestinal roundworms from pets to people. Atlanta, GA: CDC, NCID.
CDC. 1997. Guidelines for the prevention of opportunistic infections in persons infected with HIV. *MMWR* 46:4-44.
CDC. 1999. Outbreak of West Nile-Like Viral Encephalitis -- New York, 1999. *MMWR* 48(38): 845-849.
CDC. 2000(a). Preventing congenital toxoplasmosis. *MMWR* 49(2): 57-75.
CDC. 2000(b). Probable locally acquired mosquito-transmitted *Plasmodium vivax* infection—Suffolk County, New York, 1999. *JAMA* 284(4): 431-432.
CDC. 2000(c). Guidelines for preventing opportunistic infections among hematopoietic stem cell transplant recipients. *MMWR* 49(RR10): 1-128.
CDC. 2002. *HIV/AIDS Update: A Glance at the HIV Epidemic.* Atlanta, GA: CDC, NCHSTP. http://www.cdc.gov/nchstp/od/news/At-a-Glance.htm
Conti L, Lieb S, Liberti T, et al. 1995. Pet ownership among persons with AIDS in three Florida counties. *Am J Pub Health* 85: 1559-1561.
Daszek P, Cunningham AA, Hyatt AD. 2001. Anthropogenic environmental change and the emergence of infectious diseases in wildlife. *Acta Tropica* 78: 103-116.
Davies PR, Morrow WE, Deen J, et al. 1998. Seroprevalence of *Toxoplasma gondii* and *Trichinella spiralis* in finishing swine raised in different production systems in North Carolina, USA. *Prev Vet Med* 36(1):67-76.
Deutz A, Fuchs K, Auer H, Nowotny, N. 2000. Echinococcosis - an emerging disease in farmers. *N Eng J Med* 343(10): 738-739.
Dobson AP, May RM. 1986. Disease and conservation. In: Soule, M (Ed.), *Conservation Biology: the Science of Scarcity and Diversity.* MA: Sinauer Associates. pp. 345-365.
Ehnert KL, Roberto RR, Barrett L, et al. 1992. Cysticercosis: first 12 months of reporting in California. *Bulletin of the Pan American Health Organization.* 26(2):165-72.
Ericsson CD. 1998. Travelers' diarrhea: epidemiology, prevention and self-treatment. *Infect Dis Clin North Am* 12: 285-303.

Esser AC, Kantor I, Sapadin AN. 1999. Souvenir from the Hamptons - a case of cutaneous larva migrans of six months' duration. *Mount Sinai J Med 66(5-6):334-5.*
Friedmann E, Katcher AH, Lynch J, et al. 1980. Animal companions and one-year survival of patients after discharge from a coronary care unit. *Public Health Report* 95:307-312.
Furness BW, Beach MJ, Roberts JM. 2000. Giardiasis surveillance – United States, 1992-1997. *MMWR* 49(SS07): 1-13.
Gammonley, J. 1991. Pet projects. *Journal of Gerontological Nursing* 17: 12-15.
Garber LP, Salman MD, Hurd HS, et al. Potential risk factors for *Cryptosporidium* infection in dairy calves. *JAVMA 205(1):86-91.*
Gauthier JL, Richardson DJ. 2002. Knowledge and attitudes about zoonotic helminths: a survey of Connecticut pediatricians and veterinarians. *Suppl Comp Contin Educ Pract Vet* 24(4A): 10-14.
Glaser CA, Safrin S, Reingold A, Newman TB. 1998. Association between *Cryptosporidium* infection and animal exposure in HIV-infected individuals. *JAIDS* 17: 79-82.
Glickman LT, Chaudry IU, Costantino J et al. 1981. Pica patterns, toxocariasis, and elevated blood lead in children. *Am J Trop Med Hyg* 30(1): 77-80.
Glickman LT, Cypess RH. 1977. *Toxocara* infection in animal hospital employees. *Am J Public Health* 67:1193-1195.
Grant S, Olsen CW. 1999. Preventing zoonotic diseases in immunocompromised persons: the role of physicians and veterinarians. *Em Inf Dis* 5(1):159-163.
Hotez PJ. 2002. Global burden of human parasitic disease. *Comp Parasitol* 69 (2): in press.
Hotez PJ, Feng Z, Xu LQ, Chen MG et al. 1997. Emerging and reemerging helminthiases and the public health of China. *Emerg Inf Dis* 3: 303-10.
Immigration and Naturalization Service (INS). 1995. Legal Immigrations, 1990-1995. Washington, DC: Department of Justice.
Institute of Medicine Committee on Emerging Microbial Threats to Health (IMCEMTH). 1992. *Emerging Infection: Microbial Threats to Health in the United States.* National Academy Press, Washington, D.C.
Johnson DE, Miller LC, Iverson S, et al. 1992. The health of children adopted from Romania. *JAMA* 268:3446-3451.
Killick-Kendrick R. 1996. Leishmaniasis—an English disease of the future*? Bull Trop Med Int Health* 4:5.
Kramer MH, Sorhage FE, Goldstein ST, et al. 1998. First reported outbreak in the United States of cryptosporidiosis associated with a recreational lake. *Clin Inf Dis* 26(1):27-33.
Lears MK, Guth KJ, Lewandowski L. 1998. International adoption: a primer for pediatric nurses. *Pediatric Nursing.* 24(6):578-86.
Lengerich EJ, Addiss DG, Marx JJ, et al. 1993. Increased exposure to cryptosporidia among dairy farmers in Wisconsin. *J Inf Dis* 167(5):1252-1255.
Liu LX, Weller PF. 1993. Stongyloidiasis and other intestinal nematode infections. *Infect Dis Clin North Am* 7: 655-682.
Mac Kenzie WR, Hoxie NJ, Proctor ME, et al. 1994. A massive outbreak in Milwaukee of Cryptosporidium infection transmitted through the public water supply. *N Engl J Med* 331:161-7.
Marcus LC, Stambler M. 1979. Visceral larva migrans and eosinophilia in an emotionally disturbed child. *J Clin Psych* 40(3): 139-140.
McNicholas J, Collis GM. 2001. Children's representations of pets in their social networks. *Child* 27(3): 279-284.
McSweegan E. 1996. The infectious diseases impact statement: a mechanism for addressing emerging diseases. *Emerging Infectious Diseases* 2(2): 103-108.
Miller LC, Hendrie NW. 2000. Health of children adopted from China. *Pediatrics* 105(6):E76.
Melson GF, Schwarz RL, Beck AM. 1997. Importance of companion animals in children's lives – implications for veterinary practice. *JAVMA* 211(12):1512-1518.

Morin CA, Roberts CL, Mshar P, et al. 1997. What do physicians known about cryptosporidiosis? A survey of Connecticut physicians. *Arch Intern Med* 157(9): 1017.
Nowotny N, Deutz A. 2000. Letters: Preventing zoonotic diseases in immunocompromised persons: the role of physicians and veterinarians. *Emerg Inf Dis* 6(2): 208-209.
Okhuysen PC. 2001. Traveler's diarrhea due to intestinal protozoa. *Clin Inf Dis* 33: 110-114.
Olson ME, Guselle NJ, O'Handley RM, et al. 1997. *Giardia* and *Cryptosporidium* in dairy calves in British Columbia. *Canadian Veterinary Journal* 38: 703-706.
Olson ME. 2002. *Giardia* and giardiasis: a zoonotic threat. *Suppl Comp Contin Educ Pract Vet* 24(4)A: 10-14.
Paci, E. 1995. Exploring new tourism marketing opportunities around the world. In: Proceedings of the 11th General Assembly of the World Tourism Organization, Cairo, Egypt.
Park SY, Glaser C, Murray WJ, et al. 2000. Raccoon roundworm (*Baylisascaris procyonis*) encephalitis: case report and field investigation. *Pediatrics* 106(4): 56-66.
Reade, LS. 1995. Pet ownership, social support and one year survival among post myocardial patients in the CAST. *SCAS Journal* 3: 20-24.
Roberts LS, Janovy J. 1996. *Foundations of Parasitology*, 5th ed. Guilford, CT: Wm. C. Brown Publishers.
Rogers DJ, Randolph SE. 2000 The global spread of malaria in a future, warmer world. *Science*. 289(5485):1763-1766.
Rogers J, Redding J. 2000. Watch out for screwworm. APHIS Press Release. Washington, D.C.: USDA.
Sachs WJ, Adair R, Kirchner V. 2000. Enteric parasites in east African immigrants. *Minnesota Medicine*. 83(12):25-8.
Saeed AM, Harris NV, DiGiacomo RF. 1993. The role of exposure to animals in the etiology of *Campylobacter jejuni/coli* enteritis. *Am J Epi* 137: 108-114.
Schwabe, CW. 1984. *Veterinary Medicine and Human Health*, 3rd ed. Baltimore, MD: Williams and Wilkins.
Scholtens RG, Braff EH, Healey GA, Gleason N. 1968. A case of babesiosis in man in the United States. *Am J Trop Med Hyg* 17:810-813.
Sharghi N, Schantz PM, Caramico L, et al. 2001. Environmental exposure to *Toxocara* as a possible risk factor for asthma: a clinic-based case-control study. *Clin Inf Dis* 32 (7): E111-116.
Sischo WM, Atwill ER, Lanyon LE, George J. 2000. *Cryptosporidium* on dairy farms and the role these farms may have in contaminating surface water supplies in the northeastern United States. *Prev Vet Med* 43: 253-267.
Slom TJ, Cortese MM, Gerber SI, et al. 2002. An outbreak of eosinophilic meningitis caused by *Angiostrongylus cantonensis* in travelers returning from the caribbean. *New Eng J Med* 346(9): 668-675.
Slovak AJ. 1984. Toxocaral antibodies in personnel occupationally concerned with dogs. *Br J In Med* 41:419.
Spencer L. 1992. Study explores health risks and the human animal bond. *JAVMA* 201:1669.
St Pierre LA, Kreisle RA, Beck AM. 1996. Role of veterinarians in educating immunocompromised clients on the risks and benefits of pet ownership. In: Proceedings of the Geraldine R. Dodge Foundation Gathering and Reports of 1996 Veterinary Student Fellows, Ithaca, New York.
Stehr-Green JK, Schantz PM. 1986. Trichinosis in Southeast Asian refugees in the United States. *Am J Public Health* 76: 1238-1239.
Stoll N. 1947. This wormy world. *Journal of Parasitology* 33: 1-18.
Taylor LH, Woolhouse MEJ . 2000. Zoonoses and the risk of disease emergence. In: Proceedings of the Int Conf Emerg Inf Dis, Atlanta, GA, pg. 14.
Teutsch SM, Juranek DD, Sulzer A, et al. 1979. Epidemic toxoplasmosis associated with infected cats. *New Eng J of Med* 300(13):695-699.

Thomas DR, Salmon RL, Morgan-Capner P, et al. 1999. Occupational exposure to animals and risk of zoonotic illness in a cohort of farmers, farmworkers, and their families in England. *J Ag Safety and Health* 5(4):373-382.
UNOS. 2001. More than 50,000 patients in US awaiting kidney transplantation. http://www.unos.org/Newsroom/archive_newsrelease_ 20011005_ 50000 kidney.htm
USDA. 2001. *Food and Agricultural Policy: Taking Stock for the New Century.* USDA: Washington, D.C.
van Knapen F. 2000. Control of trichinellosis by inspection and farm management practices. *Veterinary Parasitology.* 93(3-4):385-392.
Wallace MR, Rosetti RJ, Olson PE. Cats and toxoplasmosis risk in HIV-infected adults. *JAMA* 1993; 269:76-7.
Welch TP. 2000. Risk of giardiasis from consumption of wilderness water in North America: a systematic review of epidemiologic data. *Int Journal of Inf Dis* 4(2):100-103.
WHO. 2000. WHO report on global surveillance of epidemic-prone infectious diseases. WHO/CDS/CSR/ISR/2000.1 http://www.who.int/emc-documents/surveillance/whocdscsrisr20001c.html
WHO. 1999. *Tuberculosis and air travel: guidelines for prevention and control.* Geneva, Switzerland: World Health Organization.
Woodruff AW, Shah AI, Prole JH. 1978. Study of toxocaral infection in dog breeders. *Br Med J* 1:51.
Wreghitt TG, Gray JJ, Balfour AH. 1986. Problems with serological diagnosis of *Toxoplasma gondii* infections in heart transplant recipients. *J Clin Path* 39(10):1135-1139.

TOXOCARIASIS AND BAYLISASCARIASIS

Zandra Hollaway Duprey and Peter M. Schantz

Division of Parasitic Diseases, National Center for Infectious Diseases, Centers for Disease Control and Prevention, Atlanta, GA

TOXOCARIASIS

Etiologic Agent: *Toxocara canis*, less commonly *Toxocara cati*; larva migrate in many tissues, immature adult in lumen of small intestine (*T. cati*).
Clinical Manifestations: Infected people experience a spectrum of disease ranging from no symptoms to eosinophilia, visceral larval migrans, or ocular larva migrans syndromes. This disease is rarely fatal.
Complications: Long-term complications such as neurological deficits and loss of vision in the affected eye may occur.
Mode of Transmission: Humans become infected by ingestion of infective *Toxocara* eggs in contaminated soil; compulsive behaviors such as geophagia predispose to accidental ingestion. Infection may also occur by ingestion of larvae in raw liver from infected chickens, cattle, and sheep.
Diagnosis: Specific antibody assays including enzyme-linked immunosorbent assay (ELISA) or immunoblot assay.
Drug(s) of Choice: Albendazole, mebendazole
Reservoir Hosts: Dogs, cats
Control Measures: Careful personal hygiene, elimination of intestinal parasites from pets by preventive treatments, and prevention of children from playing in potentially contaminated environments all contribute to reducing risk of accidental ingestion of *Toxocara* eggs.

Taxonomy and Morphology

Toxocara spp. are helminth parasites of the Phylum Nematoda, Subclass Secernentea, Order Ascaridida, Superfamily Ascaridoidea. *Toxocara canis* adult female worms measure 6-18cm and males 4-10cm. *Toxocara cati* worms are somewhat smaller measuring 4-12cm in females and measure 3-6cm in males (Glickman and Schantz, 1981).

Life Cycle

The life cycle of *Toxocara* spp. is complex having a phase of developmental arrest in intermediate or paratenic hosts (including humans) and a complete phase within its definitive host, the canine. Generally, adult *T. canis* worms are found primarily in young puppies and lactating bitches, whereas *T. cati* adult worms parasitize kittens and queens. *Toxocara canis* is a well-adapted parasite capable of vertical transmission from the bitch to her pups transplacentally and, less commonly, by transmammary passage of larvae. Bitches often reinfect themselves while caring for their pups by ingesting larval stage worms and can then develop patent infections. Female *T. canis* worms may produce up to 200,000 eggs per day. Both young puppies and bitches pass these unembryonated eggs in their stools. Within several weeks, under optimal conditions in moist soil at temperatures >27 ^{0}C, these eggs will embryonate and develop infective larvae. *Toxocara* larvae undergo several molts from first-stage to infective third-stage larvae within the egg. When ingested by susceptible hosts, infective larvae hatch from these eggs. Because *Toxocara* eggs are resistant to environmental conditions such as freezing, moisture, and extremes in pH, these parasitic stages are capable of surviving in soil for a number of years, serving as a persistent potential source of infection. Once ingested by a susceptible final host (dogs for *T. canis* and cats for *T. cati*) the larvae hatch from the egg, penetrate the wall of the intestine, and are carried away in the systemic circulation to the liver, lungs, kidneys, or other somatic tissue where they migrate and ultimately encyst as an arrested infective larvae. Infective larvae remain in this arrested state in most adult dogs. However, arrested *T. canis* larvae are reactivated in pregnant or lactating bitches and resume development and migration resulting in intestinal infection of the bitch and may also infect the pups *in utero*. After birth of the pups infected *in utero*, the infective larvae undergo tracheal migration traveling through the circulatory system to the lungs, through the trachea to the proximal gastro-intestinal tract to mature in the small intestine of the pup. Also, some reactivated larvae may be shed in the bitch's milk and infect her nursing pups by this route. Ingestion of eggs by pups older than 5 weeks of age seldom results in patent infections. These larvae undergo somatic migration and do not reach the alimentary tract (Sprent, 1958). Furthermore, most pups infected pre-natally or early in post-natal life lose their patent infections between 3 and 6 months of age. Patent *Toxocara* infections in adult dogs due to reactivation of arrested larvae rarely occur. New infections in these dogs are believed to be largely a result of ingestion of infected rodents, birds or other paratenic (transport) hosts.

Toxocara cati has a life cycle similar to *T. canis*. Migration and arrested development in paratenic hosts is also a key feature to the life cycle (Sprent, 1956). However, transplacental transmission from queen to kittens does not occur in *T. cati*. Nursing kittens may acquire infection from their dams via the transmammary route.

Transmission of *Toxocara* to humans occurs directly through ingestion of embryonated eggs from the soil or from contaminated hands, or indirectly by eating unwashed egg-contaminated raw vegetables (Salem and Schantz, 1988). Ingestion of larvae in liver, meat, or other tissues of paratenic hosts is another mode of transmission. Humans and a wide range of paratenic hosts such as sheep, rodents, and earthworms experience somatic migration after ingesting infective embryonated eggs of both *T. canis* and *T. cati.* The larvae migrate extensively through host tissues before entering an arrested state at a particular tissue site. Neither *T. canis* nor *T. cati* appear to develop into adult worms in an unintentional host and may survive for many years in somatic tissue. Therefore, paratenic hosts do not pass *Toxocara* eggs.

History

In the first half of the 20th century, several scientists speculated that some parasites of animals might be infective for humans, i.e. zoonotic. In 1952, Beaver and colleagues described larvae of *T. canis* in the tissues of children with eosinophilia-hepatomegaly syndrome, a common clinical syndrome of previously unknown cause. Visceral larval migrans was coined to describe this syndrome. Prior to this discovery, in 1950, Wilder had described nematode invasion of the eye, and, she stated that this finding had been reported previously with some reports dating as far back as 1771.Wilder misclassified this nematode as third stage hookworm larvae but in 1956 Nichols re-examined available materials and correctly identified the larvae as *T. canis*. Covert toxocariasis is a complex of clinical manifestations associated with *Toxocara* infection in humans that have been more recently described in the literature by Taylor et al., 1988. Asymptomatic toxocariasis as its name implies lacks clinical features but diagnosis can be with laboratory findings.

Epidemiology

Toxocariasis in dogs, cats, and humans apparently occurs worldwide. Studies of infection in dogs in the US, Great Britain, Canada, Brazil, the Middle East, and several African countries documented prevalences from 2 to 90%. An US national intestinal parasite survey in shelter dogs in 1993-1994 documented a 14.5% prevalence of *T. canis* (Blagburn et al.1996), however, prevalence is considerably higher in younger dogs because of prenatal and transmammary transmission. In 1999, a survey of enteric parasitic infection in kittens in Central New York State revealed *T. cati* prevalence rates of about 33% (Spain et al., 2001). A recent report indicated that approximately 16% of pet cats in Connecticut were infected with *T. cati* (Rembiesa and Richardson, 2003).

Soil contaminated with *Toxocara* eggs has been found worldwide (Anonymous, 2000). Consequently, opportunities for environmental contamination of *Toxocara* spp are nearly ubiquitous. Contaminated areas, both private and public, include yards in country, suburban, and city homes, playgrounds, parks, sandpits, gardens, fields, beaches, and paved areas such as sidewalks and streets. Contaminated soil in public places and parks, especially urban parks, provides a potential source of infection for the non-pet-owning population. In many areas, children and pets play in these limited green spaces leading to increased opportunities for transmission. Risk to the pet-owning public is obvious as these individuals have frequent and repeated exposure to dogs and cats.

In the 1990's, studies in Germany and the UK revealed rates of contaminated soil samples as high as 87% and 66%, respectively. Other studies during this time indicated that rates of positivity in surveys in Norway (39%) and Japan (42%) were also high (Mizgzjska, 2001). Approximately 30% of soil samples from certain parks in the US contained eggs (Anon, 2000). However, soil surveys have generally employed variable and non-standardized methods, therefore proportions of positive samples in different geographical locations are not always comparable (Mizgzjska, 2001).

With regard to human infection, serologic surveys in asymptomatic children have shown a wide range of toxocaral seroprevalence in different populations, with the highest prevalence usually observed in children less than 10 years of age (Schantz, 2000). The high incidence of toxocariasis in young children is related to exposure and behavior. Young children are at a higher risk of disease because they are more likely to play in contaminated environments, engage in childhood behaviors such as pica and geophagia, or put fingers, toys, or other *Toxocara* contaminated objects in their mouths. Seroprevalance averaged 3% in the US, with rates in some subpopulations as high as 23% (Glickman and Schantz, 1981). In a more recent serosurvey of children aged 2-15 years in Connecticut the highest seroprevalence was in urban Hispanic children of the 2-10 age group (Table 1) (Sharghi et al., 2001).

As previously mentioned, urban spaces often provide opportunities for increased exposure because of the concentration of pets and children in limited green spaces. Intercity children are more likely to play in these urban spaces than their surburban counterparts. The intercity populace tends to be less educated and poorer than those that reside elsewhere. Thus, socioeconomic status can be a confounding factor.

Globally, seroprevalence varies from 0-4 % in urban Spain and Germany, 31% in Ireland, 66% in rural Spain, to 83% in some Caribbean subpopulations (Anon, 2000).

Pathogenesis

Humans acquire toxocariasis by ingesting eggs that contain infective larvae. The eggs hatch in the proximal small intestine and release larvae which penetrate the intestinal mucosa, migrate to the liver via portal circulation, and continue through the vasculature to the lungs to enter systemic circulation. *Toxocara canis* larvae are impeded when the size of the larvae exceeds the diameter of the blood vessel they occupy. When this

Table 1. Profile of Connecticut children seropositive for Toxocara.

Demographic	% of all seropositive individuals
Age	
2-10 years	54.55
11-15 years	45.45
Race	
White	15.15
Black	18.18
Puerto Rican	51.52
Other Hispanic	9.09
Other	6.06
Residence	
Suburban	23.33
Urban	73.33
Rural	3.33

From Sharghi et al. (2001)

Table 2. Toxocaral seroprevalence by geographic region (%)

Geographic location	Seroprevalence (%)
Caribbean	83
Germany, urban	0-4
Ireland	31
Spain, urban	0-4
Spain, rural	66
United States	3-23

occurs, the larvae traverse the vascular wall and migrate in the surrounding tissue, burrowing in the tissue causing hemorrhage necrosis and secondary inflammation (Taylor and Holland, 2001). Larvae migrate extensively throughout the body and have been found in every tissue and organ system, including liver, lungs, heart, and brain. Some of the larvae die whereas others become dormant for years before re-activation. Size of inoculum and frequency of reinfection determine the distribution and survival of larvae and the clinical severity of the disease. Acute *T. canis* larvae infection induces a Th2-type CD4+ cellular immune response with marked IgE production and eosinophilia, neutrophilia, and monocytosis (Liu, 2001). This delayed hypersensitivity reaction is thought to be dose-dependent. Chronic infection leads to granuloma formation around the larvae containing eosinophils, multi-nucleated giant cells, and fibrous tissue. Although most commonly found in the liver, larval granulomas can also be found in other tissue such as the lungs, central nervous system (CNS) and ocular tissue. *Toxocara* larvae continue to survive and migrate despite the host's vigorous immunologic response. This seems to be associated with the larvae's ability to shield themselves from host immunity by producing and shedding substances from the larval surface (Smith et al., 1981).

Clinical Manifestations and Complications

Asymptomatic toxocariasis, visceral larval migrans (VLM), ocular larval migrans (OLM), and the lesser-known covert toxocariasis are all clinical manifestation of infection with *Toxocara* parasites. Multiple factors determine clinical manifestation of toxocariasis: the number of eggs ingested, chronicity of infection, migration patterns of the larvae, host immunity, and other poorly understood factors. The clinical manifestations of toxocariasis are dose dependent. Ingestion of lower doses of infective larvae leads to the development of asymptomatic eosinophilia, covert toxocariasis, or ocular larval migrans. Ingestion of larger and/or repeated doses of infective larvae may result in the VLM syndromes. Rarely, intense acute infection can lead to VLM with concurrent OLM.

Visceral Larval Migrans

Visceral larval migrans, usually diagnosed in young children (average age 2 years), is marked by the inflammatory response to numerous larvae migrating in liver and other tissues. This clinical manifestation of toxocariasis is characterized by persistent eosinophilia, anemia, leukocytosis, fever, hepatomegaly, hypergammaglobulinemia, positive titers to *Toxocara* larval antigens and elevated titers of blood group isohemaglutinins. Clinical signs often include wheezing or coughing; pulmonary infiltration, such as bronchopneumonia is evident in one-third of patients. Recurrent bronchitis has also been noted in *Toxocara* seropositive patients. Some authors have suggested a positive relationship between toxocariasis and asthma, however conflicting reports indicate that the evidence is nonconclusive (Sharghi et al., 2000).

Because the liver is the most frequently involved organ, hepatomegaly is a common finding. Hepatic lesions can present as space-occupying masses that resemble metastatic cancer, however, any organ can be affected. Occasionally, splenomegaly or lymphadenopathy is evident. Integumentary manifestations may include urticaria, migratory cutaneous lesions or nodules (Taylor and Holland, 2001). Rheumatologic abnormalities can include arthralgias, and vasculitis. Rarely, *T. canis* has been associated with eosinophilic myositis, panniculitis and cardiac pseudotumor with sudden death. Neurologic manifestations, including focal or generalized seizures, muscular paresis and coma, have been reported.

Ocular Larval Migrans

The epidemiologic profile of patients with OLM is different from that of patients with VLM. Ocular involvement is more common in older children and adults; hence most OLM sufferers lack the associated recent histories of puppy contact, pica, geophagia, and lack signs and symptoms of VLM. It is believed that with fewer infecting larvae, the immune response is less and the eye may be invaded randomly several months after infection.(Schantz, 1989). Larval invasion of the eye is usually unilateral and common symptoms include visual loss, strabismus due to disuse of the affected eye, and more rarely eye pain (Pollard et al., 1979). Squinting in the affected eye may be noted. Routine examination may reveal poor or absent sight in one eye.

Examination of the fundus may reveal a retinal scar or, if invasion of the eye has been recent, a tumor formed by the inflammatory granuloma (Taylor and Holland, 2001). Some vitreous haze is common. Occasionally, a cataract develops. Other reported presentations of OLM are uveitis, vitreous abscess, pars planitis, optic neuritis, hypopyon and isolated peripheral granulomas, and chronic endophthalmitis resulting in retinal detachment . A tumor-like lesion of the fundus of one eye and peripheral eosinophilia are suggestive of intra-ocular toxocariasis. Rarely, periorbital swelling caused by a subcutaneous granuloma containing toxocaral larva is a clinical finding. As previously mentioned, concomitant OLM and VLM syndrome rarely occurs.

Covert Toxocariasis

The term covert toxocariasis was suggested by Taylor et al., 1988 to describe the signs and symptoms in patients with clinical features that are singly nonspecific but together form a recognizable symptom complex. When associated with elevated toxocaral antibody titers, signs and symptoms including abdominal pain, anorexia, sleep and behavioral disturbances, cervical adenitis, wheezing, limb pain, headache, failure to thrive in children, and fever is suggestive of covert toxocariasis. (Taylor and Holland, 2001)

Differential Diagnosis

Toxocaral VLM must be differentiated from signs and symptoms caused by other tissue-migrating helminths (ascarids, hookworm, filariae, *Strongyloides stercoralis*, and *Trichinella spiralis*) and hypereosinophilic syndromes. Hepatic capillariasis can be confused with toxocariasis involving the liver; patients have hepatomegaly, eosinophilia, and abnormal results of liver function tests. A liver biopsy may demonstrate eggs of *Capillaria* species within a hepatic larval granuloma.

Ocular disease may be confused with retinoblastoma, ocular tumors, developmental anomalies, exudative retinitis (Coats disease), trauma, and other childhood uveitides.

Diagnostic Tests

The diagnosis of toxocariasis should be considered for any person with persistent eosinophilia. The clinical and laboratory findings other than specific serologic tests are nonspecific and do not always differentiate toxocaral larval migrans from other conditions resulting in eosinophila. Because the larval ascarids do not develop further in human tissues, adult worms do not reach the intestinal lumen and diagnosis by egg detection is not possible.

Ocular larval migrans should be considered in any patient with raised, unilateral, whitish, or gray lesions in the fundus or endopthalmitis (Pollard et al., 1979). Such ocular lesions and an elevated antibody titer to *T. canis* are suggestive of ocular toxocariasis. In patients with ocular lesions compatible with OLM, positive ELISA titers support the diagnosis but do not rule out retinoblastoma or other possible conditions. Ophthalmic lesions have also been characterized using computed tomography and ultrasound imaging. Demonstration of specific toxocaral antibody reactivity in aqueous or vitreous fluid strongly supports the diagnosis (Pollard et al, 1979).

Histologic examination of biopsy tissue may identify larvae within infected tissue or granulomas. However, biopsy is often unrewarding as specimens may not yield larvae, unless the infection is massive. Other diagnostic tools such as ultrasound, computed tomography, and magnetic resonance imaging may be helpful in the diagnosis of toxocariasis. Focal lesions in the liver and nervous tissue have been reported using these imaging modalities. (Duprez et al., 1996).

Enzyme-linked immunosorbent assay (ELISA) or immunoblot assays, using excretory-secretory antigens from infective-stage larvae, provide the best serologic diagnosis for VLM and OLM (Glickman and Schantz, 1986). In patients whose clinical signs and history suggest VLM, positive toxocaral ELISA titers are strong presumptive evidence of toxocaral infection. A rising or falling titer of at least two-fold, in a recently ill patient, is consistent with VLM. However, because toxocaral antibody titers may remain elevated for years after infection, a measurable titer is not proof of a causative relationship between *T. canis* and the patient's current illness. Toxocaral ELISA testing is available at the CDC and several other diagnostic laboratories in the US.

Treatment

Asymptomatic toxocariasis does not require anthelmintic treatment. Although there are isolated reports of ocular disease occurring years after an episode of VLM, available data suggest that nearly all asymptomatic seropositive individuals have spontaneous resolution of their eosinophilia and seroreactivity without adverse sequelae. However, severe VLM is a potentially fatal condition because of the possibility of larval migration causing hemorrhage, necrosis, and secondary inflammation in the brain, lungs, and heart therefore corticosteroid therapy may be warranted in conjunction with anthelmintics.

Treatment of patients with VLM is primarily supportive. Anthelmintics such as albendazole (400mg PO bid for 5 days) and mebendazole (100-200mg PO bid for 5 days) can be instituted (Anon, 2002). Older drugs such as diethylcarbamazine (50 to 150mg PO tid for 1 to 3 weeks) and thiabendazole(25 to 50 mg/kg/day PO for 1 to 3 weeks) have been effective in the management of VLM, albeit inconsistently, and have smaller margins of safety than albendazole and mebendazole.

Treatment of acute ocular toxocariasis is directed toward suppressing the inflammatory response associated with larval migration or worm death. Systemic and intraocular anti-inflammatories (prednisone 30 mg to 60 mg PO each day for 2 to 4 weeks; triamcinolone acetonide 40 mg sub-Tenon weekly for 2 weeks; or topical prednisiolone acetate) are the most consistently effective form of intervention if commenced within the first 4 weeks of illness. Concurrent use of anthelmintic drugs has not provided additive benefit in managing OLM. Laser photocoagulation is another method of treatment by destroying the migrating larvae, if visualization of the nematode is possible.

Treatment of patients with long-standing ocular *Toxocara* infection is also problematic. Corticosteroids may be effective in treating exacerbations of ocular inflammation, although relapse and progression of ocular disease is common. Intraocular adhesions, retinal traction and detachment, retrolental plaques, and chronic vitreal inflammation are most effectively managed by surgical intervention. Surgical procedures most commonly used include par plana vitrectomy and scoleral buckling.

Prevention

Most cases of human toxocariasis can be prevented by simple measures, such as careful personal hygiene, not allowing children to play in potentially contaminated environments, and the elimination of parasites from pets. Increased public awareness regarding potential zoonotic hazards and how to minimize them is crucial, especially among pet owners. Veterinarians are uniquely suited to provide pet owners with sound advice (CDC/AAVP, 1995). They have the knowledge and have established rapport with clients, and a high proportion of pet owners use veterinary services. Unfortunately, repeated surveys of pet owners reveal that few people (laymen as well as pediatricians!) are aware of the potential health hazards associated with pets (Gauthier and Richardson, 2002).

All pet dogs and cats should receive anthelmintics periodically to prevent morbidity caused by maternally-acquired infection and to prevent dissemination of infectious eggs. It is particularly important to treat puppies and bitches shortly after whelping to prevent passage of infectious eggs (CDC/AAVP, 1995). Most pups are born infected with *T. canis*, therefore it is recommended to treat pups beginning at 2 weeks of age before eggs are first passed in the feces. Additional treatment should be administered at 4, 6, and 8 weeks. Such protocol should be followed to kill worms originating from transplacental, transmammary and fecal-oral routes of transmission (Sherding and Johnson, 1994). Several effective anthelmintics are available for treatment of roundworms in pets (CDC/AAVP, 1995). Pyrantel pamoate (10mg/kg PO if greater than 2.5kg or 5mg/kg PO if less than 2.5kg) is a cost-effective, efficacious, safe treatment of toxocariasis in puppies, kittens and older dogs and cats.

Recent advances in veterinary pharmaceutics have simplified intestinal parasite control in pets. A variety of currently available anthelmintics used for prophylaxis of heartworm infection (*Dirofilaria immitis*) are also effective for eliminating ascarids and hookworms. When used as recommended, these agents provide effective suppression and elimination of intestinal nematode infections (CDC/AAVP, 1995).

Contemporary Challenges

The crux of prevention of toxocariasis is prudent deworming of dogs and cats and prevention of environmental toxocaral contamination. Veterinarians are essential stewards of public health and zoonosis control in this respect. However, veterinarians face challenges when recommending adequate anthelmintic protocol to their clients. Most veterinarians have initial contact with puppies at 6-8 weeks of age or older. At this point, infected puppies will have patent infections, possibly having contaminated the home environment with *T. canis* eggs. More conscientious and knowledgeable breeders will have an established deworming protocol, but the layperson may not be aware of their role in prevention of this zoonosis. In addition, prescription anthelmintics administered by veterinarians versus less expensive "over-the-counter" self-administered anthelminthic preparations adds to the confusion over effective and cost-effective treatment of pets. Guidelines for effective treatment of dogs and cats and counseling of pet owners regarding the potential zoonotic transmission of ascarids and hookworms are provided in the CDC/AAVP Recommendation for Veterinarians.

Public parks and public areas can provide opportunity for toxocaral infection if clean-up ordinances and leash laws are not observed. Laws and ordinances in public parks and public areas should be more readily enforced to decrease disease transmission. Finally, parents should not allow children, especially those with pica and geophagia, to play outdoors where they are likely to have access to infectious eggs.

BAYLISASCARIASIS

Etiologic Agent: *Baylisascaris procyonis*
Clinical Manifestations: Moderate to severely infected people may experience nausea, lethargy, ataxia, decreased cognitive awareness, paresis, fever, photophobia, blindness, or death. Mildly infected people may be asymptomatic or show few signs of disease.
Complications: Profound neurologic disease, ocular damage or blindness can be sequelae to these infections.
Mode of Transmission: Humans can become infected by ingesting eggs from soil, water, contaminated hands or contaminated fomites, such as toys.
Diagnosis: Serologic testing including ELISA and IFA. Demonstration of specific *Baylisascaris* antibodies in the cerebrospinal fluid.
Drugs of Choice: No effective, curative treatment available. However, immediate post-exposure albendazole or mebendazole with concurrent corticosteroid use may be effective. Anthelmintics, corticosteroids, and other symptomatic treatments have been unsuccessful in patients having infections of long duration.
Reservoir Hosts: Predominantly raccoons; skunks and badgers are hosts to other potentially zoonotic species of *Baylisascaris, B. columnaris* and *B. melis*, respectively.
Control Measures: Careful personal hygiene. Children with pica behavior or geophagia should be carefully supervised when playing outdoors. Discourage feeding of raccoons near homes or children's playgrounds.

Taxonomy and Morphology

Baylisascaris spp. are classified as Phylum Nematoda, Order Ascaridida, Superfamily Ascaridoidea). These tan-colored adult worms can reach a size of 20-22cm long in the female and 9-11 cm long in the male (Kazacos, 2001).

Life Cycle

The raccoon is the natural definitive host of *B. procyonis* (Kazacos, 2001). Infection in the raccoon is generally subclinical and adult worms reside in the small intestine (Kazacos, 2001). Raccoons usually acquire infection through ingestion of paratenic hosts. Infected raccoons can shed millions of eggs in their feces daily as adult female *B. procyonis* worms are extremely prolific (Kazacos, 2001). These eggs may remain viable for years in moist soil. They are extremely resistant to environmental degradation and attempted decontamination. *B. procyonis* larvae undergo several molts within eggs in the environment and with suitable temperature, reach the infective stage in approximately 2-4 weeks. (Kazacos, 2001)

When ingested by an accidental avian or mammalian host, the eggs hatch and larvae are released. Larvae then penetrate the gut wall and migrate into various tissues. This aggressive and widespread migration can lead to clinical VLM, OLM, or NLM (neural larval migrans) in a variety of paratenic hosts including dogs, rabbits, birds, and humans (Kazacos, 2001). Pet dogs infected this way can harbor patent infections and shed eggs into the environment as well.

Like *Toxocara spp.*, transmission of *B. procyonis* to humans occurs directly through ingestion of embryonated eggs from the soil or indirectly from contaminated hands or toys. Eggs of *B. procyonis* tend to be sticky and adhere well to surfaces and objects (Kazacos, 2002).

History

The first confirmed cases of baylisascariasis in humans were described in 1984 and 1985 in two boys: a 10-month old from Pennsylvania and a 18-month old from Illinois (Huff et al, 1984, Fox et al, 1985). These boys presented with a rapidly progessive and ultimately fatal eosinophilic meningoencephalitis (Kazacos, 2002). An earlier unconfirmed case of NLM was documented in an 18-month old girl in 1975 from Missouri, who presented with clinical signs and laboratory findings consistent with baylisascariasis such as acute hemiplagia, CSF and peripheral eosinophilia and elevated isohemagglutinns (Anderson et al, 1975). Existing technology at that time was not able to differentiate this infection from other larval migrans syndromes, therefore the diagnosis was presumptive based upon these clinical signs and laboratory findings (Kazacos, 2002).

Epidemiology

Baylisascaris spp. are found in several sylvatic hosts including raccoons, skunks, and bears. While many species of this parasite are capable of infecting humans, *B. procyonis* is most devastating in the paratenic host. *Baylisascaris procyonis* is endemic in North American raccoons with prevalence ranging from 68-82% in raccoon populations in the Midwest, Northeast, and West Coast (Kazacos 2001) (Figure 1).

Soil contamination may be widespread as raccoons are common throughout the US in rural, suburban, and urban areas. In areas where these well-adapted creatures are densely populated, soil contamination with infective eggs may also be concentrated. Raccoon defecation habits contribute to high densities of these eggs in the soil. "Latrines," the preferred

Figure 1. Geographic occurrence of *Baylisascaris procyonis* infection in raccoons in the USA.*

*Shaded states indicate positive reports of intestinal *Baylisascaris procyonis* in raccoons according to published reports cited in Kazacos (2001). Data is approximate and limited by the absence or limitations of published surveys in some states.

communal sites where raccoons defecate, can be found most often in the base of trees, raised flat surfaces such as large logs, old tree stumps, wooden decks, patios, and rooftops (Kazacos, 2001, Page et al., 1999).

Baylisascariasis is becoming increasingly recognized for its role as a cause of severe and fatal NLM in young patients. Localization apparently restricted to the eye, or OLM, is more frequently diagnosed in older patients. Since 1981, at least 12 cases of severe or fatal encephalitis resulting from infection with *B. procyonis* have been identified in the US in California, Illinois, Michigan, Minnesota, New York, Oregon, and Pennsylvania (Kazacos, 2002). As with toxocariasis, children are most often infected due to behavioral characteristics such as geophagia and pica. Thus, children are at a higher risk of acquiring the infection. All of the documented cases of baylisascariasis have been in boys, although this predominance is more likely a function of the play-habits of boys rather than an inherent increased susceptibility to infection (Gavin, 2002). Of the 12 cases reported since 1981, 10 of the 12 occurred in children aged 9 months to 6 years. Eight of these children were less than 19 months of age (Kazacos, 2002). Epidemiological studies of confirmed cases suggest that infants and young children, especially boys, with pica and geophagia and contact with infected raccoons or their feces are at highest risk for CNS disease (Gavin, 2002).

Pathogenesis

When ingested, eggs hatch in the intestine of the paratenic host, penetrate the gut wall, and are carried through the body by the circulatory system. (Kazacos, 2001). Pathogenesis is similar to that of *Toxocara* infection , but the disease is usually more severe. Migration of *B. procyonis* is more aggressive than *Toxocara* spp., often involving the central nervous system. Unlike *Toxocara* spp., *B. procyonis* larvae grow while migrating (Kazacos, 2001). The combination of its migratory behavior and larval growth ensures that, if the inoculum is sufficient, the likelihood of severe neurologic disease is very high. Neurologic disease is further complicated by delayed encapsulation of *B. procyonis* larvae within the CNS as compared to other tissues (Kazacos, 1997). In addition to growth and persistence of the larval nematode in the CNS, the migrating larvae secrete and excrete toxic products while migrating. This elicits a marked peripheral and cerebrospinal fluid (CSF) eosinophilic inflammatory response. These released cytotoxic eosinophilic granule proteins and larval products are neurotoxic (Moertel et al., 2001, Kazacos, 1997, Hamann et al, 1989). Therefore, the delayed and relatively ineffective host inflammatory response probably contributes to the severe encephalitis and neuroretinitis that often occurs in baylisascariasis (Moertel et al., 2001).

Clinical Manifestation

The clinical manifestations of human baylisascariasis are mild to severe. As in toxocariasis, the severity of clinical disease is dependent upon dose and migration patterns of the larvae, especially the number of larvae entering the CNS or eye (Kazacos and Boyce, 1989, Kazacos, 2000, Kazacos, 2001). The most severe manifestation of baylisascariasis is acute, fulminant NLM which is usually seen in individuals who have apparently ingested large numbers of *B. procyonis* eggs (Kazacos, 2000). Those with lighter infections usually only suffer milder clinical disease or may be asymptomatic (Gavin, 2002). Larvae can also migrate throughout the body resulting in fever, hepatomegaly and respiratory disease.

Ocular larval migrans without other clinical manifestation may be a sequela to ingestion of smaller numbers of larvae. Baylisascariasis has been implicated in diffuse unilateral subacute neuroretinitis (Goldberg, 1993). Ocular larval migrans may also be seen in conjunction with NLM as a result of larval dissemination associated with heavy infection (Kazacos, 2001, Rowley et al, 2000, Park et al , 2000).

Diagnosis

Encephalopathy with or without retinitis in addition to peripheral and CSF eosinophilia with radiographic evidence of diffuse white matter disease strongly suggests *B. procyonis* NLM (Kazacos, 2002). Unfortunately, the diagnosis of baylisascariasis is often late as it is usually only considered in the presence of eosinophilic meningoencepahlitis, a stage of clinical disease in which major neurologic damage has already been sustained. Positive tests for specific antibodies in CSF or serum confirm the diagnosis of baylisascariasis (Kazacos, 2002). Elevated serum isohemaglutinins may also suggest this diagnosis. As in toxocariasis, larval stages do not reach the intestine and develop to the adult worm therefore fecal demonstration of the eggs or larvae is unlikely. Diagnostic imaging modalities such as CT or MRI may reveal lesions ranging from nonspecific abnormalities of cerebellar and periventricular deep white matter to global atrophy (Cunningham et al., 1994; Moertel et al., 2001). However, changes visualized by neuroimaging lags behind severe clinical manifestations (Kazacos, 2002).

Treatment

The prognosis of *B. procyonis* NLM is grave with or without treatment (Kazacos, 2000, Kazacos, 2001). Albendazole, mebendazole, thiabendazole, levamisole, and ivermectin have been unsuccessful in the treatment of NLM (Anonymous, 2002) possibly because the diagnosis of the helminth occurred late in the disease process (Kazacos, 2002). Survivors of the disease may have profound neurologic complications despite treatment.

Several anthelmintics, such as piperazine citrate (120-240 mg/kg), pyrantel pamoate (6-10mg/kg), and fenbendazole (50-100mg/kg x 3-5 days) have been effective in treating intestinal *B. procyonis* infection in raccoons (Kazacos, 2001).

Prevention

Prevention of baylisascariasis is critical as there are no viable treatment options for this disease. Education of the public is the most important aspect in preventing human infection. Pediatricians, veterinarians, and public health officials should be at the forefront of public education and awareness of the inherent danger with raccoon feces.

Good hygiene and discouragement of raccoon ownership (pet raccoons) or feeding of wild raccoons is helpful in avoiding risk of baylisascariasis (Kazacos, 2002). If parents are aware of raccoon latrines in the environment they should be properly cleaned up and children should be prohibited from playing in these areas. Pica and geophagia should also be discouraged and special care to avoid contaminated areas must be implemented for children exhibiting these behaviors. The importance of hand washing after outdoor play or animal contact should be stressed (Kazacos 2000, Kazacos 2001). Individuals involved in wildlife rescue and rehabilitation should be made aware of this infection and counseled to observe appropriate precautions when handling raccoons and their cage environments.

Decontamination of areas contaminated with *B. procyonis* eggs is difficult. These eggs are resistant to decontamination with common disinfectants including bleach and can survive in the environment for years (Kazacos, 2001).Use of a 1:1 xylene:ethanol solution can be effective although often impractical, in decontaminating areas after organic debris has been removed (Kazacos and Boyce, 1989). Heat is the best method to kill the eggs and can be accomplished by flaming the area (including soil) or burning affected material such as straw or wood cages (Kazacos, 2001). Boiling water, steam cleaners, or autoclaves are also effective methods of decontaminating areas (Kazacos, 2001). All cages, carriers, or areas (regardless of material) occupied by raccoons should be thoroughly cleaned. Other animals should not be housed in cages previously occupied by raccoons as they could acquire the infection, which can lead to profound CNS disease and death. When cleaning raccoon latrines, care should be taken to prevent accidental ingestion of eggs. Personal protection equipment such as disposable coveralls, rubber gloves, washable rubber boots, and a particulate face mask should be used to prevent the inhalation or ingestion of any eggs or fecal fungi stirred up while cleaning (Kazacos and Boyce 1989). Once cleanup has been completed all disposable equipment should be incinerated, autoclaved, or otherwise properly disposed of (Kazacos, 2001).

REFERENCES

Anderson DC, Greenwood R, Fishman M, et al. 1975. Acute infantile hemiplasia with cerebrospinal fluid eosinophilic pleocytosis: An unusual case of visceral larva migrans. *J Pediatr* 86: 247-249.

Anonymous. 2000. Toxocariasis. In: Chin, James (Eds.) *Control of Communicable Diseases Manual 17th ed.* American Public Health Association. pp. 497-499.

Anonymous. 2002. Drug for Parasitic Infections. *The Medical Letter*, January 2002; 1-12.

Blagburn BL, Lindsay DS, Vaughan JL et al. 1996. Prevalence of canine parasites based on fecal flotation. Comp Contin Educ Vet Prac 18:483-509.

CDC. 1995. *Recommendations for veterinarians; How to Prevent Transmission of Intestinal Roundworms in Pets and People.* Atlanta, Georgia: National Center for Infectious Disease, Centers for Disease Control and Prevention. 8pp.

Cunningham CK, Kazacos KR, McMillian JA, et al. 1994. Diagnosis and management of *Baylisascaris procyonis* infection in an infant with nonfatal meningoencephalitis. *Clin Infect Dis* 18: 868-872.

Duprez, TP, Bigaignon G, Delgrange, E, et al. 1996. MRI of cervical cord lesions and their resolution in *Toxocara canis* myelopathy. *Neuroradiology* 38 (8): 792-795.

Fox AS, Kazacos KR, Gould, NS et al. 1985. Fatal eosinophilic meningoencephalitis and visceral larval migrans caused by the raccoon ascarid *Baylisascaris procyonis. N Engl J Med* 312: 1619-1623.

Gauthier JL, Richardson DJ. 2002. Knowledge and attitudes about zoonotic helminths: A survey of Connecticut pediatricians and veterinarians. *Suppl Comp Contin Eudc Prac Vet* 24(4A): 10-14.

Glickman LT and Schantz PM. 1981. Epidemiology and pathogenesis of zoonotic toxocariasis. Epidemiol Rev 3: 230-250.

Glickman LT, Schantz PM, Grieve RB. 1986. Toxocariasis. In: Walls KW, Schantz PM (Eds.), *Immunodiagnosis of Parasitic Diseases, Vol.1. Helminthic Diseases*. New York: Academic Press. pp. 201-231.
Goldberg MA, Kazacos KR, Boyce WM, et al. 1993. Diffuse unilateral subacute neuroretinitis. Morphometric, serologic, and epidemiologic support for *Baylisascaris* as a causative agent. *Ophthal* 100(11): 1695-1701.
Hamann KJ, Kephart GM, Kazocos KR, Gleich GJ. 1989. Immunofluorescent localization of eosinophil granule major basic protein in fatal human cases of *Baylisascaris procyonis* infection. *Amer J Trop Med Hyg* 40: 291-297.
Huff DS, Neafie RC, Binder MJ et al. 1984. The first fatal *Baylisascaris* infection in humans: An infant with eosinophilic meningioencephalitis. *Pediatr Pathol* 2: 345-352.
Kazacos, KR. 2001. *Baylisascaris procyonis* and related species. In: Samuels WM, Pybus MJ, Kocan AA (Eds.), *Parasitic diseases of wild mammals. 2nd ed.* Ames, Iowa: Iowa State University Press. pp. 301-341.
Kazacos KR. 2000. Protecting children from helminthic zoonoses. *Contemp Pediatr* 17 (Supplement): 1-24.
Kazacos KR. 1997. Visceral, ocular, and neural larval migrans. In: Connor DH, Chandler FW, Schwartz, DA, et al. (Eds.), *Pathology of infectious diseases, Vol. II.* Stamford, Connecticut: Appleton and Lange, pp. 1459-1473.
Kazacos KR, Boyce WM. 1989. *Baylisascaris* larval migrans. *J Am Vet Med Assoc* 195:894-903.
Kazacos KR, Gavin PJ, Shulman ST, et al. 2000. Raccoon roundworm encephalitis: Chicago, Illinois, and Los Angeles, California, 2000. *MMWR* 50 (51): 1153-1155:
Liu, LX. 2001. Toxocariasis and Larval Migrans Syndromes. In: Guerrant RG, Walker DH, Weller PF (Eds.), *Essentials of Tropical Infectiuous Diseases*. Philadelphia, Pennsylvania: Churchill Livingstone. pp.428-433.
Mizgajska H. 2001. Eggs of *Toxocara* spp. in the environment and their public health implications. *J Helminthol* 75:147-151
Moertel CL, Kazacos KR, Butterfield JH, et al. 2001. Eosionphil-associated inflammation and elaboration of eosinophil-derived proteins in two children with raccoon roundworms (*Baylisascaris procyonis*) encephalitis. *Pediatrics* 2001: 108.
Page LK, Swihart RK, Kazacos KR. 1999. Implications of raccoon latrines in the epizootiology of baylisascariasis. *J Wild Dis* 35:474-480.
Park SY, Glaser C, Murray WJ, et al. 2000. Raccoon Roundworm (*Baylisascaris procyonis*) encephalitis: Case report and field investigation. *Pediatrics* 2000;106.
Pollard ZA, Jarrett WH & Hagler WS. 1979. ELISA for diagnosis of ocular toxocariasis. *Ophthalmology* 86: 743-9.
Rembiesa C, Richardson DJ. 2003. Helminth Parasites of the house cat, *Felis catus*, in Connecticut, USA. *Comp Parasitol* 70: In press.
Rowley HA, Uht RM, Kazazos KR, et al. 2000. Radiologic-pathologic findings in raccoon roundworm (*Baylisascaris procyonis*) encephalitis. *Amer J Neuroradiol* 21: 415-420.
Salem G, Schantz PM. Toxocaral visceral larva migrans after ingestion of raw lamb liver. *Clin Inf Dis* 1992; 15: 743-744.
Schantz PM. 1989. Toxocara larva migrans now. Am J Trop Med Hyg 41:21-34.
Schantz PM. 2000. Toxocariasis. In: Strickland GT (Ed.), *Hunter's Tropical Medicine and Emerging Infectious Diseases, 8th ed.* Philadelphia: W. B. Saunders Company. pp. 787-790.
Sharghi N, Schantz PM, Hotez P. 2000. Toxocariasis: An occult cause of childhood neuropsychological deficits and asthma? Sem Ped Inf Dis 11:257-260.

Sharghi N, Schantz PM, Caramico L et al. 2001. Environmental exposure to *Toxocara* as a possible risk factor for asthma: A clinic-based case-control study. *Clin Inf Dis.* 32: 111-116.
Sherding RG, Johnson SE. 1994. Diseases of the Intestines. In: Birchard SJ, Sherding RG (Eds.), *Saunders Manual of Small Animal Practice.* Philadelphia: W. B. Saunders Company. pp. 695-697.

Smith HV, Quinn R, Kusel JR, Girdwood RWA. 1981. The effect of temperature and antimetabolites on antibody binding to the outer surface of second stage *Toxocara canis* larvae. *Mol. Biochem Parasitol* 2: 183-193.
Spain CV, Scarlett JM, Wade SE, et al. 2001. Prevalence of enteric zoonotic agents in cats less than 1 year old in central New York state. *J Vet Intern Med* 15 (1): 33-38.
Sprent JFA. 1956. The life history and development of *Toxocara cati* (Schrank, 1788) in the domestic cat. *Parasitol* 48:184-209.
Sprent JFA. 1958. Observations on the development of *Toxocara canis* (Werner, 1782). *Parasitol* 48:185-198.
Taylor MRH, Keane CT, O'Connor P, et al. 1988. The expanded spectrum of toxocaral disease. Lancet 1:692-694.
Taylor MRH, Holland C. 2001. Toxocariasis. In: Gillespie SH, Pearson RD (Eds.), *Principles and Practice of Clinical Parasitology*. New York: John Wiley & Sons, LTD. pp. 501-519.

TRICHINELLOSIS

Dickson D. Despommier

The Joseph L. Mailman School of Public Health, Columbia University, New York, New York

Etiologic Agents: *Trichinella spiralis, T. britovi, T. nelsoni, T. nativa, T. pseudospiralis*

Clinical Manifestations: The early stages of the infection (week 1-3 post-infection) in the small intestine cause nausea, diarrhea, and vomiting. Systemic disease (week 3-12 post-infection) results from migration of newborn larvae to the tissues of all organs. Worms penetrate cells, killing them. Heart and brain tissue is most sensitive to invasion, and in moderate to heavy infection, a constellation of signs and symptoms often mimicking other diseases is produced, including photophobia, confusion, coma, fever, muscle pain, oedema (bilateral periorbital), splinter hemorrhages, conjunctivitis, and cardiac arythmias.

Complications: Heart failure, coma, death.

Mode of Transmission: Ingestion of raw or undercooked meats (striated skeletal muscle tissue) containing infective larvae.

Diagnosis: Definitive: microscopic identification of larvae in biopsy of striated skeletal muscle tissue, antigen capture ELISA, PCR. Indirect: positive serology (IgG ELISA); prolonged (2-5 weeks) elevated circulating eosinophilia, bilateral peri-orbital edema, history of eating raw or undercooked meat (pork, beef, lamb, wild game animals) within the last month prior to onset of clinical signs and symptoms.

Reservoir Hosts: Nearly all mammals, even "strict" herbivores, are susceptible to infection, but scavenger species (rats, raccoons, hyenas, etc.) are the dominant reservoir hosts in natural ecological settings. *Trichinella spiralis* is worldwide in distribution and occurs most frequently in domestic pigs. *Trichinella britovi* and *T. nativa* are common among wild animals of the arctic and sylvan environments. *Trichinella nelsoni* is found exclusively in Africa among a wide variety of carnivores. *Trichinella pseudospiralis* is a parasite of raptors, but a few human cases have been reported. *Trichinella papuae* is a newly described species, and its host range has yet to be fully described. It infects wild and domestic pigs, and human cases have been reported in Papua, New Guinea.

Control Measures: Meat inspection at the slaughterhouse is the surest way to prevent infection in humans. Cooking any meat at 137^0 F for 10 minutes kills infective larvae of all trichinella species. Freezing meat until solidly frozen kills larvae of *T. spiralis*, but not larvae of *T. britovi* and *T. nativa*.

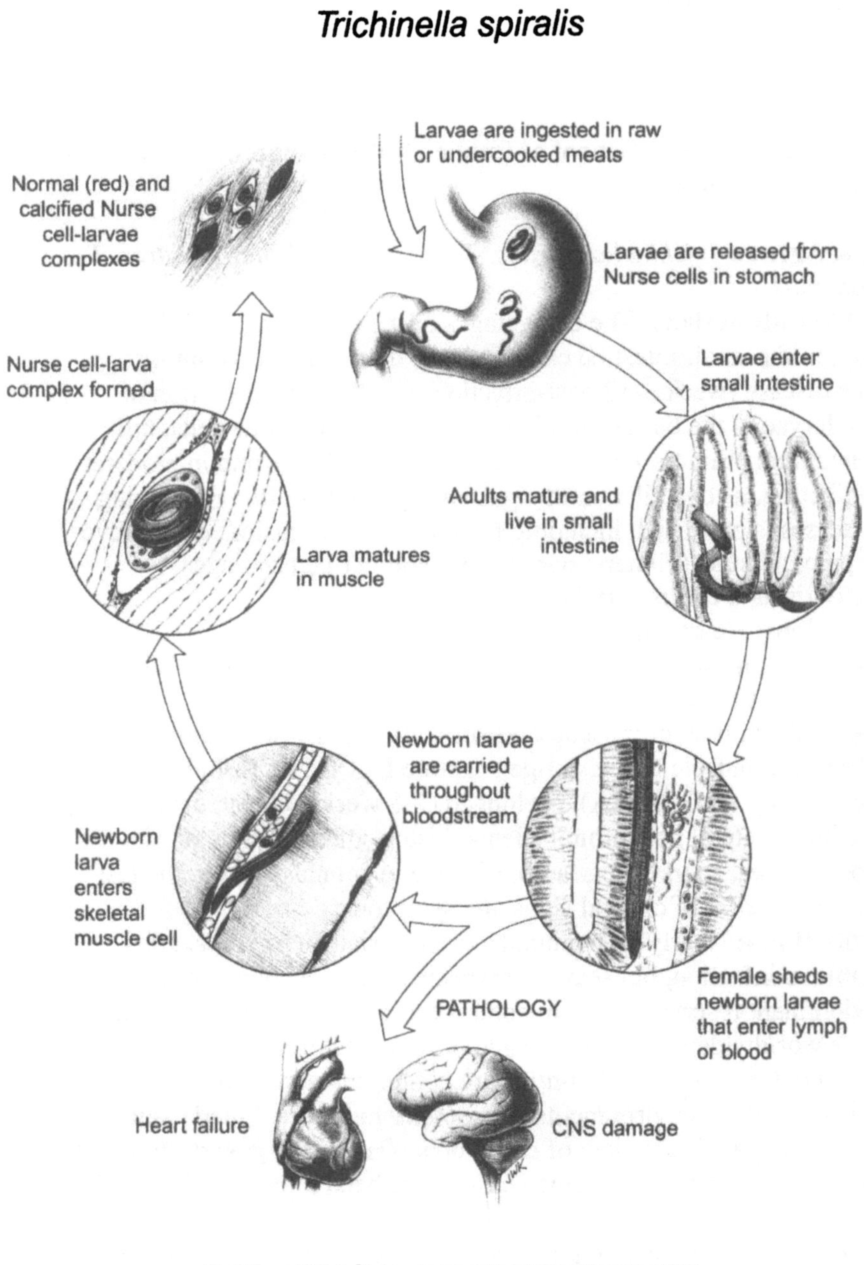

Figure 1. Life cycle of Trichinella spiralis. *Drawing by John Karapelou.*

Trichinella spp. are parasitic nematodes infecting a wide range of mammals, making them one of the most widely distributed parasite groups in the world. Six species (Zarlenga et al., 1999; Pozio, 2001; Tibayrenc, 2001) in the Order Trichurida have been described: *Trichinella spiralis, T. nativa, T. britovi, T. nelsoni, T. pseudospiralis*, and *T. papuae*. All are intracellular parasites at some time during their life cycle (Despommier, 1983). The diseases they cause are referred to collectively as trichinellosis. Most human infections are caused by a single species, *Trichinella spiralis*, followed in importance by *T. nativa* and *T. britovi.* Domestic pigs are the dominant reservoir host for *T. spiralis*. Its prevalence is significantly higher in some parts of Europe, Asia, and South East Asia than in the United States (Moorehead et al., 1999). It is also endemic in China and Japan (Gasser et al., 1998; Yiman et al., 2001). The major reservoir hosts for *T. nativa* are polar bears and walruses (Serhir et al., 2001). *Trichinella britovi* is also sylvatic, being found in Asia and Europe, infecting numerous carnivorous animals (e.g., fox, wild boar, opossum, dog, cat). *Trichinella pseudospiralis* infects mainly birds, and rarely humans (Jongwutiwe et al., 1998). *Trichinella papuae* has been isolated from wild and domestic pigs (Pozio et al., 1999) and from humans (Owen et al., 2001) in Papua, New Guinea. *Trichinella nelsoni* occurs in Equatorial Africa, with hyenas and large cats serving as reservoir hosts (Pozio et al., 1991; Pozio et al., 1997).

NATURAL HISTORY

Only the life cycle of *T. spiralis* will be described (Figure 1), since it causes most of the clinical cases, worldwide. The biological pattern is similar for *T. nativa, T. britovi,* and *T nelsoni.* In contrast, *T. pseudospiralis* and *T. papuae* differ somewhat from these species in the muscle phase of their life cycle, because they fail to form a fully developed Nurse cell (Evensen et al., 1989; Pozio, 2001). Infection begins with the ingestion of raw or undercooked meats containing the Nurse cell-larva complex. Larvae (1mm x 36 μ; Figure 2) are

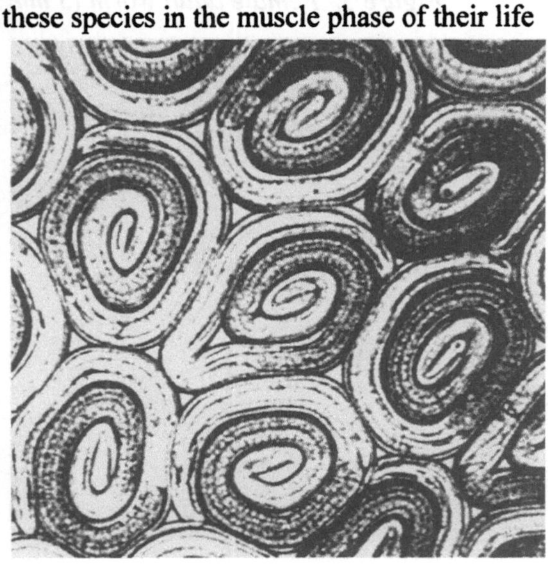

Figure 2. L1 larvae freed from Nurse cells (each larva measures 1 mm x 36 μm).

freed by digestive enzymes in the stomach, and rapidly locate to the upper two-thirds of the small intestine. The worms' epicuticle (Kennedy et al., 1987) is altered by alkaline pH and host digestive enzymes This allows environmental cues to be received by the parasite that alter its behavior, and aid it in selecting a site within columnar epithelium (Stewart et al., 1987). The worm penetrates a row of cells at the base of the villus. At this point, they are intra-multi-cellular organisms.

Larvae molt four times in rapid succession over a 30-hour period, developing to sexually mature adults (males are 1.5mm x 36 μm; females are 3mm x 36 μm, (Figure 3). Mating immediately ensues. Patency occurs five days later. Females give birth to live offspring, referred to as newborn larvae. The female produces offspring as long as host acquired immunity does not interfere. Newborn (i.e., migratory) larvae gain entrance to the lamina propria, and using their spear-like stylet, enter lymphatic vessels or capillaries. They eventually enter the general circulation and become distributed throughout the body.

Newborns emerge from the capillary beds of various organs and penetrate cells. They may either stay or leave, depending upon the nature of the environmental cues received from the host cell. Most cell types in the body die as the result of invasion. The striated skeletal muscle cell is the lone notable exception (Figure 4). Several days after entering a muscle cell, the larva stimulates a portion of the cytoplasm adjacent to it to de-differentiate, then to re-differentiate into a new entity that proceeds to function by supporting the growth and development of the parasite (Figure 5). This

Figure 3. Female adult worm (3 mm X 36 μm).

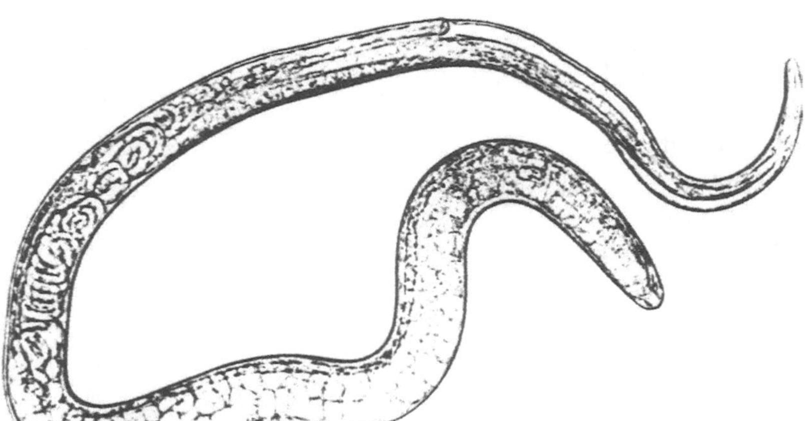

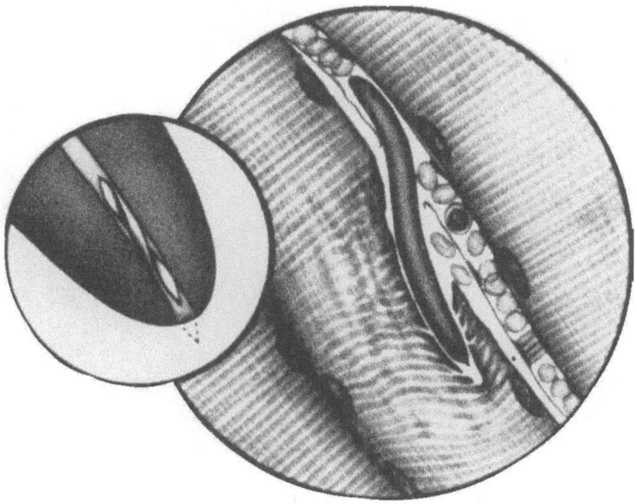

Figure 4. Newborn larva penetrating muscle cell. Drawing by John Karapelou.

process is termed Nurse cell formation (Despommier, 1998), and takes 20 days to complete (Despommier et al., 1975). The Nurse cell-larva complex can remain intact for long periods of time, up to the life span of the infected host, in some cases. In nature, this remarkable adaptation to parasitic life enables *Trichinella* to become widely dispersed among the wildlife of a given region. Larval development within the Nurse cell, however, is precocious. Parasites that are 14 days old or older are now infectious for the next host.

Worms that penetrate other cell types fail to induce Nurse cell formation, and either re-enter capillaries, or become surrounded by granulomas and eventually die. The enteral phase can last for up to 2-3 weeks, but eventually all adult worms are expelled by acquired immune responses, mediated by IL-4 (Urban et al., 2001), IL-5 (Vallance et al., 2000), IgE, and eosinophils (Bell, 1998). Host immunity is elicited by a plethora of molecules consisting, in part, of the highly antigenic tyvelose sugar moiety (Dea-Ayuela et al., 2001; Goyal et al., 2002) secreted by the developing parasite in the small intestine (Gagliardo et al., 2002).

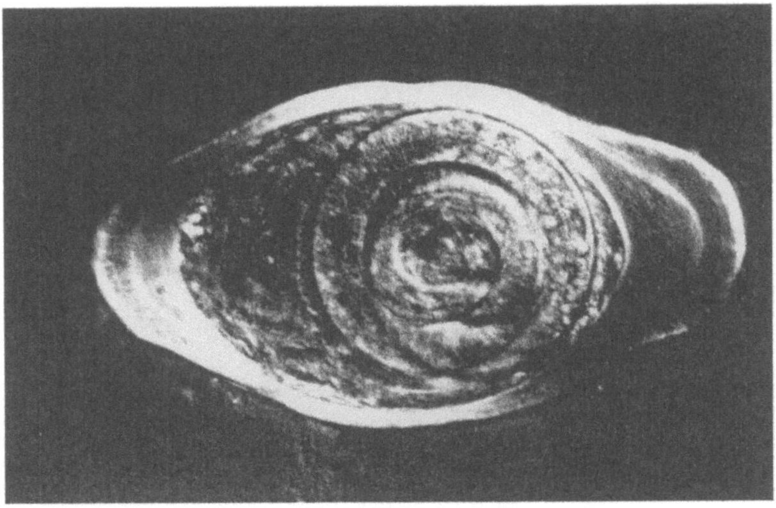

Figure 5. Nurse cell-parasite complex. Photograph by Eric Grave.

EPIDEMIOLOGY

Outbreaks of trichinellosis typically occur sporadically (Pozio, 2000), and are almost invariably traceable back to a common source of infected meat. Hunters sharing their kill (Nutter et al., 1998), and domestic pigs are the usual sources of infection (Moorehead et al., 1999). Infection is more likely in older populations living in endemic regions where ingestion of raw or undercooked pork or meats, prepared similarly from other domesticated species of mammals, is traditional. Mexico (Ortiega-Pierres et al., 2000), Central and Eastern Europe (Seroka, 2001; Olteanu, 2001; Marinculic et al., 2001), Egypt (Morsy et al., 2000), most countries of South America (Ortega-Pierres et al., 2000; Schenone et al., 1997; Bartoloni et al., 1999), Thailand (Takahashi et al., 2000), Japan (Kudo et al., 2001; Yamaguchi, 1991), China (Wang and Cui, 2001), and Lebanon (Blondheim et al., 1984; Haim et al., 1997) have reported outbreaks of trichinellosis in recent years. Within the United States, only a few cases are reported each year (Moorehead et al., 1999).

PATHOGENESIS

During the enteral infection, the developing parasites damage columnar epithelium. A secretory enteritis results. As adult worms begin shedding offspring some 5 days later, a local inflammation, consisting of eosinophils, neutrophils, plasma cells, and lymphocytes, develops adjacent to the parasites and intensifies. A bacteremia due to enteric flora may result from large numbers of newborns entering the lamina propria, and in severe infections, death may ensue from septicemia. In experimental infections in immunologically defined strains of laboratory animals, the number of larvae produced was dependent upon numerous factors related to the immune capabilities of a given strain (Despommier, 1983). In rodents, interleukin 4 (Urban et al., 2001) and 9 (Faulkner, 1997), eosinophils, and IgE antibodies are effector mechanisms for expelling adults (Finkelman et al., 1997; Lawrence et al., 1998). Whether or not these same mechanisms occur during human infection is not known.

One of the most interesting features of the parenteral phase is the Nurse cell-larva complex, enabling the parasite to remain in the host for long periods of time. Its formation is characterized by a remarkable series of cellular and molecular re-arrangements, all of which are parasite-directed (Despommier, 1998). Myofilaments and other muscle cell-specific structures are replaced over 14-16 days with smooth membranes and aerobically dysfunctional mitochondria. The nuclei adjacent to the developing parasite enlarge and divide, amplifying the host's genome within the Nurse cell cytoplasm (Despommier et al., 1991; Jasmer, 1993; Yao et al., 1998). Over-expression of collagen type IV and type VI mRNA results in the production of an outer acellular capsule (Polvere et al., 1997). Angiogenesis is induced, driven by VEGF synthesis, beginning on day 7 after the larva enters the muscle cell (Capo et al., 1998). A circulatory rete is formed (Baruch and Despommier, 1991), surrounding the capsule. It most likely functions to allow the worm to obtain nutrients and dispose of its wastes (Figure 6). Nurse cell formation is induced by the secretions of the developing parasite (Despommier, 1998), which emanate from the stichosome (Despommier and Muller, 1976; Takahashi and Araki, 1991). Some of them enter the host cell nuclei and remain there throughout the infection (Despommier et al., 1990), possibly functioning as transcription factors (Despommier, 1998; Yao and Jasmer, 1998; Yao and Jasmer, 2001).

As larvae penetrate cells, they cause extensive damage. This situation is most problematic to the host when it involves the heart and central nervous system. Myocarditis is typical in heavy infections, and is transitory, since Nurse cells cannot form in heart muscle cells (Ursell et al., 1984). In the central nervous system, newborn larvae tend to wander before leaving the tissue. Much of the inflammation in the brain is initiated by petechial hemorrhages. After parasites enter skeletal muscle fibers, they induce an infiltration of mixed inflammatory cells. Edema and myositis develop about 14 days after the penetration of muscle fibers. The extent of damage is primarily related to the number of larvae produced by adults in the small intestine. Most Nurse cell-parasite complexes become calcified and die within months after forming.

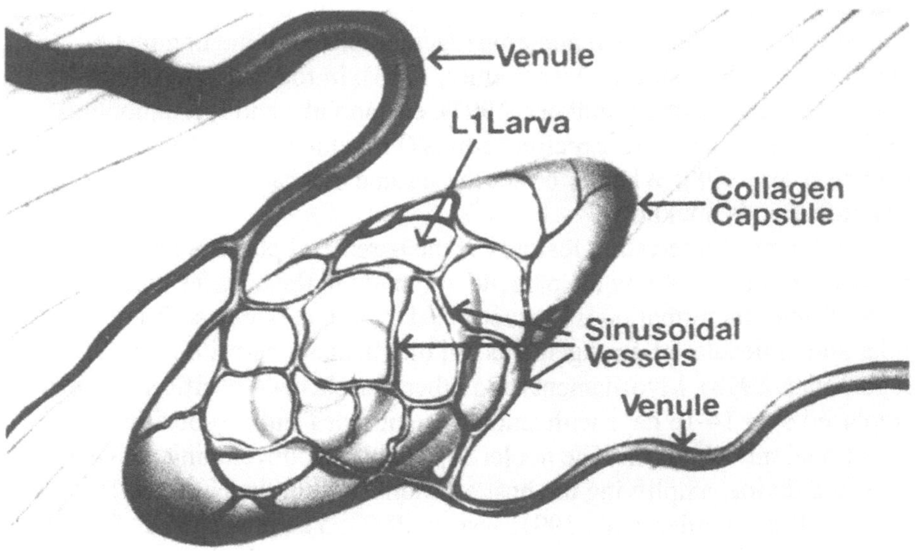

Figure 6. Nurse cell-parasite complex. Drawing by John Karapelou.

CLINCAL MANIFESTATIONS AND COMPLICATIONS

Trichinellosis resembles a wide variety of clinical conditions, and for that reason it is often misdiagnosed (Murrell and Bruschi, 1994; Capo and Despommier, 1996; Clausen et al., 1996; Kociecka, 2000). The severity of signs and symptoms is dose dependent. There are, however, important clinical clues in the early stages of the disease that should lead the physician to include trichinellosis into the differential diagnosis (Figure 7). The first few days of the infection are characterized by a gastroenteritis associated with diarrhea, abdominal pain, and vomiting. This phase is transitory, and abates within 10 days. A history of eating raw or undercooked meat helps to rule in this parasitic infection at this time. Others who also ate the same meat, and are suffering similarly reinforces the suspicion of trichinellosis. Unfortunately, most clinicians favor a diagnosis of food poisoning at this juncture in the infection.

The parenteral phase begins approximately 1 week after infection and may last several weeks. Typically, the patient's diarrhea abates, followed by fever and myalgia, bilateral periorbital edema, and petechial hemorrhages seen most clearly in the subungual skin, but also observed in the conjunctivae and mucous membranes. Muscle tenderness is a common complaint. Laboratory studies reveal an elevated white blood cell count (12,000-15,000 cells/mm^3), and a circulating eosinophilia ranging from 5% to 50% (Venturiello et al., 1995). The diagnosis of food poisoning is replaced with numerous other plausible diagnoses. But again, trichinellosis usually is given a low priority status on that list, too.

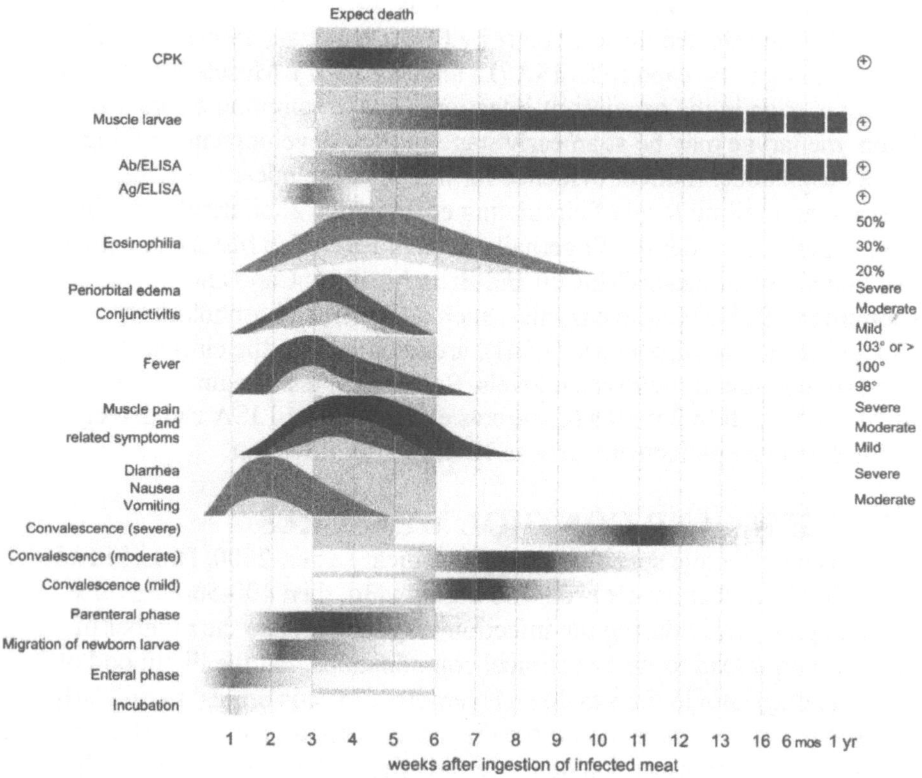

Figure 7. Summary of clinical symptoms and signs.

Larvae penetrating tissues other than muscle gives rise to more serious symptoms and signs. In many cases of moderate to severe infection, cardiovascular involvement leads to myocarditis. Electrocardiographic (ECG) changes are frequently noted during this phase. Parasite invasion of the diaphragm and the accessory muscles of respiration results in dyspnia. Neurotrichinellosis also occurs in association with central nervous system invasion (Hess et al., 1982; Taratuto and Venturiello, 1997; Nikolic et al., 1998). The convalescent phase follows the acute phase, usually uneventfully.

Two clinical presentations have been described for *T. nativa* infections resulting from the ingestion of infected polar bear or walrus meat: a classic myopathic form (see above), and a second form that presents as a persistent diarrheal illness (MacLean et al., 1989). The second form is thought to represent a secondary infection in previously sensitized individuals.

DIAGNOSTIC TESTS

Definitive diagnosis depends on finding the Nurse cell-parasite complex by microscopic examination of a muscle biopsy, detection of *Trichinella*–specific DNA in infected muscle tissue by PCR (Wu et al., 1999), or finding circulating antigen by capture ELISA (Li and Ko, 2001). Muscle biopsy can be negative, even in the heaviest of infections, due to sampling errors. In addition, the larvae may be at an early stage of their development, making them inconspicuous. Indirect evidence for infection includes: 1. a rising, plateauing, and falling level of circulating eosinophils, 2. bilateral periorbital edema, 3. petichae under the fingernails, 4. high fever, 5. a history of eating raw or undercooked meats (Murrell and Bruschi, 1994; Capo and Despommier, 1996). Muscle enzymes, such as creatine phosphokinase (CPK) and lactic dehydrogenase (LDH), are released into the circulation causing an increase in their serum levels. Serological tests begin to show positive results within 2 weeks (Contreras et al., 1999). ELISA can detect antibodies in some patients as early as 12 days after infection.

TREATMENT AND CONTROL MEASURES

There is no effective specific therapy (Medical Letter, 2000; Bruschi and Murrell, 2002). Mebendazole (200-400 mg tid x 3d, then 400-500 mg tid x 10d) when given early during the infection may help reduce the number of larvae that might lead to further clinical complications, but the likelihood of making the diagnosis in time to do so is remote. Corticosteroids, particularly prednisolone, are recommended if the diagnosis is secure. Because of their potent immunosuppressive potential, steroids should be administered with caution. Steroids must also be given when administering mebendazole. Destroying larvae without employing steroids may exacerbate host inflammatory responses and worsen disease (e.g., Jarisch-Herxheimer reaction). The parenteral phase is treated supportively with antipyretics and analgesics (aspirin, acetaminophen), until the fever and allergic signs recede.

Control measures include meat inspection at the slaughterhouse, and monitoring feed for domestic pigs for the presence of unprocessed or raw meat scraps. No meat products should be fed to pigs as nutritional supplementation, especially with the advent of bovine spongiform encephalitis. Nonetheless, some unscrupulous farmers still feed fresh meat scraps, obtained at slaughter, to pigs resulting in herd infections. Countries participating in the Common Market have instituted regulations that ban such practices. China and the United States have yet to institute on-line slaughterhouse inspection systems. Yet, despite the lack of this public health practice, the incidence of trichinellosis in the United States remains low. Its near eradication has most likely been as the result of a gradual, overall improvement in sanitary practices, both on the farm and at the abattoir (Gamble et al., 2001). United States Department of Agriculture regulations prohibiting the use of uncooked meat scraps obtained at slaughter as food for pigs undoubtedly also helps the vast majority of commercial pig farms remain free of trichinella infection. Sporadic outbreaks due to unsupervised, locally raised and slaughtered pigs remains a problem. Educating hunters about the sylvatic nature of the infection (LeCount and Zimmermann, 1986; Schad et al., 1986; Dworkin et al., 1996) could help to further reduce health risks associated with eating meat from wild animals.

Cooking all meats to 137 ^{0}F for 10 minutes renders them harmless with respect to *T. spiralis.* Freezing meat until solid also kills the larvae. *Trichinella nativa* and *T. britovi* larvae are not killed by freezing (Kapel et al., 1999).

RECENT ADVANCES AND CONTEMPORARY CHALLENGES

"Data mining" parasite genomes should be included into the mix of the next wave of pharmacological discovery. With respect to trichinella, deciphering the mechanisms by which this parasite engineers its Nurse cell comes up high on a list of laboratory-based research projects aimed at deciphering the molecular communications system that this fascinating parasite has evolved in order for it to be able to live a long-term life in its host. Parasite-specific immunosuppressive molecules, transcription factors, and inhibitors of aerobic mitochondrial activity all must surely be encoded within its genome. By subtractive hybridization with *Caenorhabditis elegans*, it should be possible to select most genes of interest from *Trichinella spiralis* that enable it to carry out its life as a parasite. Among them will be the genes for Nurse cell up-regulation.

Determining how the circulatory rete that surrounds the Nurse cell-parasite complex is induced and maintained could lead to the discovery of novel proteins that could then be applied in medical situations unrelated to clinical trichinellosis. The parasite induces VEGF synthesis via its stichosome secretions. However, unlike normal angiogenic responses to wound healing or tumor growth, for example, parasite-directed VEGF mRNA synthesis apparently does not involve the induction of hypoxia inducible factor 1. Furthermore, sinusoid-like vessels are elicited, not capillaries. Biologically active trichinella secreted proteins, once isolated and characterized, could thus prove useful in combating diseases in which the synthesis of sinusoids is one of the desired outcomes. Isolating and characterizing parasite-specific substances from trichinella might yield molecules whose activities include prevention of allograft rejection, quelling of various inflammatory responses, and the induction of angiogenesis, to name but a few.

REFERENCES

Bartoloni A, Cancrini G, Bartalesi F, et al. 1999. Antibodies against *Trichinella spiralis* in the rural population of the Province of Cordillera, Bolivia 5: 97-99.

Baruch AM, Despommier DD. 1991. Blood vessels in *Trichinella spiralis* infections: a study using vascular casts. *J Parasitol* 77(1): 99-103.

Bell RG. 1998. The generation and expression of immunity to *Trichinella spiralis* in laboratory rodents. *Advances in Parasitol* 41: 149-217.

Blondheim DS, Klein R, Ben-Dror G et al. 1984. Trichinosis in southern Lebanon. *Isr J Med Sci* 20:141-4.

Bruschi F, Murrell KD. 2002. New aspects of human trichinellosis: the impact of new Trichinella species. *Postgrad Med J* 78:15-22.

Capo V, Despommier DD. 1996. Clinical aspects of infection with *Trichinella spp. Clin Microbiol* Rev 47-54.

Capo VA, Despommier DD, Polvere RI. 1998. *Trichinella spiralis*: vascular endothelial growth factor is up-regulated within the nurse cell during the early phase of its formation. *J Parasitol* 84: 209-214.

Clausen MR, Meyer CN, Kratz T, et al. 1996. *Trichinella* infection and clinical disease. *QJM* 89: 631-636.

Contreras MC, Acevedo E, Aguilera S, et al. 1999. Standardization of ELISA IgM and IgA for immunodiagnosis of human trichinosis. *Bol Chil Parasitol* 54: 104-109.

Dea-Ayuela MA, Ubeira FM, Pitarch A, et al. 2001. A comparison of antigenic peptides in muscle larvae of several Trichinella species by two-dimensional western-blot analysis with monoclonal antibodies. *Parasite* 8 (2 Suppl):S117-119.

Despommier DD, Aron L, Turgeon L. 1975. *Trichinella spiralis*: Growth of the intracellular (muscle) larva. *Exp Parasitol* 37: 108-116.

Despommier DD, Muller M. 1976. The stichosome and its secretion granules in the mature muscle larva of *Trichinella spiralis. J Parasitol* 62: 775-85.

Despommier DD. 1983. Biology. In: Campbell WC (Ed.), *Trichinella and trichinellosis.* New York: Plenum Press, Pubs. pp. 75-152.

Despommier DD, Gold AM, Buck SW, et al. 1990. *Trichinella spiralis*: secreted antigen of the infective L1 larva localizes to the cytoplasm and nucleoplasm of infected host cells. *Exp Parasitol* 71: 27-38.

Despommier DD, Symmans WF, Dell R. 1991. Changes in Nurse cell nuclei during synchronous infection with *Trichinella spiralis. J Parasitol* 77: 290-295.

Despommier DD. 1998. How does trichinella make itself a home? *Parasitol Today.* 318-323.

Dworkin MS, Gamble HR, Zarlenga DS, et al. 1996. Outbreak of trichinellosis associated with eating cougar jerky. *J Infect Dis* 174: 663-666.

Evensen O, Skjerve E, Bratberg B. 1989. Myofibre changes and capsule formation in mice infected with different strains of *Trichinella. Acta Vet Scand* 30: 341-346.

Faulkner H, Humphreys N, Renauld JC, et al. 1997. Interleukin-9 is involved in host protective immunity to intestinal nematode infection. *Eur J Immunol* 27: 2536-2540.

Finkelman FD, Shea-Donohue T, Goldhill J et al. 1997. Cytokine regulation of host defense against parasitic gastrointestinal nematodes: lessons from studies with rodent models. *Annu Rev Immunol* 15: 505-533.

Gagliardo LF, McVay CS, Appleton, JA. 2002. Molting, ecdysis, and reproduction of *Trichinella spiralis* are supported in vitro by intestinal epithelial cells. *Infect Immun* 70: 1853-1859.

Gamble HR, Pyburn D, Anderson LA, et al. 2001. Verification of good production practices that reduce the risk of exposure of pigs to *Trichinella. Parasite* 8(2 Suppl): S233-235.

Gasser RB, Zhu XQ, Monti JR, et al. 1998. PCR-SSCP of rDNA for the identification of *Trichinella* isolates from mainland China. *Mol Cell Probes.* 12: 27-34.

Goyal PK, Wheatcroft J, Wakelin D. 2002. Tyvelose and protective responses to the intestinal stages of *Trichinella spiralis. Parasitol Int* 51:91-98.

Haim M, Efrat M, Wilson M, et al. 1997. An outbreak of *Trichinella spiralis* infection in southern Lebanon. *Epidemiol Infec* 119: 357-362.
Hess B, Frei F, Kummer H et al. 1982. Trichinellosis with neurological complications. Case report and short overview. *Schweiz Med Wochenschr* 112: 1145-1151.
Jasmer DP. 1993. *Trichinella spiralis* infected skeletal muscle cells arrest in G2/M and cease muscle gene expression. *J Cell Biol* 121: 785-793.
Jongwutiwes S, Chantachum N, Kraivichian P, et al. 1998. First outbreak of human trichinellosis caused by *Trichinella pseudospiralis*. *Clin Infect Dis*. 26:111-115.
Kapel CM, Pozio E, Sacchi L, et al. 1999. Freeze tolerance, morphology, and RAPD-PCR identification of *Trichinella nativa* in naturally infected arctic foxes. *Parasitol* 85:144-147.
Kennedy MW, Foley M, Kuo YM, et al. 1987. Biophysical properties of the surface lipid of parasitic nematodes. *Mol Biochem Parasitol* 22: 233-240.
Kociecka W. 2000. Trichinellosis: human disease, diagnosis and treatment. *Vet Parasitol* 93: 365-383.
Kudo N, Arima R, Ohtsuki M, et al. 2001. The first host record of trichinosis in a red fox, *Vulpes vulpes japonica*, from Aomori Prefecture, northern Honshu, Japan. *J Vet Med Sci* 63: 823-826.
Lawrence CE, Paterson JC, Higgins LM, et al. 1998. IL-4-regulated enteropathy in an intestinal nematode infection. *Eur J Immunol* 28: 2672-2684.
LeCount AL, Zimmermann WJ. 1986. Trichinosis in mountain lions in Arizona. *J Wildl Dis* 22: 432-434.
Li CK, Ko RC. 2001. The detection and occurrence of circulating antigens of *Trichinella spiralis* during worm development. *Parasitol Res* 87: 155-162.
MacLean JD, Vaillet J, Law C, et al. 1989. Trichinosis in the Canadian Arctic: report of five outbreaks and a new clinical syndrome. *J Infect Dis* 160: 513-520.
Marinculic A, Gaspar A, Durakovic E, et al. 2001. Epidemiology of swine trichinellosis in the Republic of Croatia. *Parasite* 8(2 Suppl): S92-94.
Medical Letter. 2000. The medical letter. New Rochelle, New York: The Medical Letter, Inc. March 2000.
Moorehead A, Grunenwald PE, Dietz VJ, Schantz PM. 1999. Trichinellosis in the United States, 1991-1996: declining but not gone. *Am J Trop Med Hyg*. 60: 66-69.
Morsy TA, Ibrihim BB, Haridy FM, et al. 2000.*Trichinella* encysted larvae in slaughtered pigs in Cairo (1995-1999). *J Egypt Soc Parasitol* 30: 753-760.
Murrell D, Bruschi F. 1994. Clinical trichinellosis. In: Sun T (Ed.), *Progress in Clinical Parasitology*. Boca Raton, FL: CRC Press. pp. 117-150.
Nikolic S, Vujosevic M, Sasic M, et al. 1998. Neurologic manifestations in trichinosis. *Srp Arh Celok Lek* 126: 209-213.
Nutter FB, Levine JF, Stoskopf, et al. 1998. Seroprevalence of *Toxoplasma gondii* and *Trichinella spiralis* in North Carolina black bears (*Ursus americanus*). *J Parasitol* 84: 1048-1050.
Olteanu G. 2001. Trichinellosis in Romania: a short review over the past twenty years. *Parasite* 8(2 Suppl): S98-99.
Ortiega-Pierres MG, Arriaga C, Yepez-Mulia L. 2000. Epidemiology of trichinellosis in Mexico, Central and South America. *Vet Parasitol* 93: 201-225.
Owen IL, Pozio E, Tamburrini A, et al. 2001. Focus of human trichinellosis in Papua New Guinea. *Am J Trop Med Hyg* 65:553-557.
Polvere RI, Kabbash CA, Capo CA, et al. 1997. *Trichinella spiralis*: synthesis of type IV and type VI collagen during nurse cell formation. *Exp Parasitol* 86: 191-199.
Pozio E, La Rosa G,. Verster A. 1991. Identification by isoenzyme patterns of two gene pools of *Trichinella nelsoni* in Africa. *Ann Trop Med Parasitol* 85: 281-283.
Pozio E, DeMeneghi D, Roelke-Parker ME, et al. 1997. *Trichinella nelsoni* in carnivores from the Serengeti ecosystem, Tanzania. *J Parasitol* 83: 1195-1198

Pozio E., Owen IL, La Rosa G, et al. 1999. *Trichinella papuae* n.sp. (Nematoda), a new non-encapsulated species from domestic and sylvatic swine. *Int J Parasitol* 29: 1825-39.
Pozio, E. 2000. Factors affecting the flow among domestic, synanthropic and sylvatic cycles of *Trichinella*. *Vet Parasitol* 93: 241-262.
Pozio E. 2001. New patterns of *Trichinella* infection. *Vet Parasitol* 98: 133-148
Schad GA, Leiby DA, Duffy CH, et al. 1986. *Trichinella spiralis* in the black bear (*Ursus americanus*) of Pennsylvania: distribution, prevalence and intensity of infection. *J Wildl Dis* 22: 36-41.
Schenone H, Lopez R, Barilari E et al. 1997. Current trends of the epidemiology of human trichinosis in Chile. *Bol Chil Parasil* 52: 22-25.
Serhir B, MacLean JD, Healey S, et al. 2001. Outbreak of trichinellosis associated with arctic walruses in northern Canada, 1999. *Can Commun Dis Rep 2001* 27: 31-36.
Seroka D. 2001. Trichinelliasis in Poland 1999. *Przegl Epidemiol* 55: 155-158.
Stewart GL, Despommier DD, Burnham J, et al. 1987. *Trichinella spiralis*: behavior, structure, and biochemistry of larvae following exposure to components of the host enteric environment. *Exp Parasitol* 63: 195-204.
Takahashi Y, Araki T. 1991. Morphological study of the stichocyte granules of *Trichinella spiralis* muscle larvae. *Jpn J Parasitol* 38: 77-85.
Takahashi Y, Mingyaun L, Waikagul J. 2000. Epidemiology of trichinellosis in Asia and the Pacific Rim. *Vet Parasitol* 93: 227-239.
Taratuto AL, Venturiello SM. 1997. Trichinosis. *Brain Pathol* 7: 663-672.
Tibayrenc, M. 2001. The relevance of evolutionary genetics for identification of *Trichinella* sp. and other pathogens at the strain, subspecies and species levels. *Parasite* 8(2 Suppl):S21-23.
Urban JF Jr., Noben-Trauth N, Schopf L, et al. 2001. Cutting edge: IL-4 receptor expression by non-bone marrow-derived cells is required to expel gastrointestinal nematode parasites. *J Immunol* 167: 6078-6081.
Ursell PC, Habib A, Babchick O, et al. 1984. Myocarditis caused by *Trichinella spiralis*. *Arch Pathol Lab Med* 108: 4-5.
Vallance BA, Matthaei KI, Sanovic S, et al. 2000. Interleukin-5 deficient mice exhibit impaired host defense against challenge *Trichinella spiralis* infections. *Parasite Immunol* 22: 487-492.
Venturiello SM, Giambartolomei GH, Constantino SN. 1995. Immune cytotoxic activity of human eosinophils against *Trichinella spiralis* newborn larvae. *Parasite Immunol* 17:555-559.
Wang ZQ, Cui J. 2001. Epidemiology of swine trichinellosis in China. *Parasite* 8(2 Suppl): S67-70.
Wu Z, Nagano I, Pozio E, Takahashi Y. 1999. Polymerase chain reaction-restriction fragment length polymorphism (PCR-RFLP) for the identification of *Trichinella* isolates. *Parasitology* 118 (Pt 2): 211-218.
Yamaguchi T. 1991. Present status of trichinellosis in Japan. *Southeast Asian J Trop Med Public Health* 22 (Suppl): 295-301.
Yao C, Jasmer DP. 1998. Nuclear antigens in *Trichinella spiralis* infected muscle cells: nuclear extraction, compartmentalization and complex formation. *Mol Biochem Parasitol* 92: 207-218.
Yao C, Bohnet S, Jasmer DP. 1998. Host nuclear abnormalities and depletion of nuclear antigens induced in *Trichinella spiralis*-infected muscle cells by the anthelmintic mebendazole. *Mol Biochem Parasitol* 96: 1-13.
Yao C, Jasmer DP. 2001. *Trichinella spiralis*-infected muscle cells: abundant RNA polymerase II in nuclear speckle domains co-localizes with nuclear antigens. *Infect Immun* 69: 4065-4071.
Yiman AE, Nonaka N, Sakai H, et al. 2001. First report of *Trichinella nativa* in red foxes (*Vulpes vulpes schrencki*) from Otaru City, Hokkaido, Japan. *Parasitol Int* 50: 121-127.

Zarlenga DS, Chute MB, Martin A, Kapel CM. 1999. A multiplex PCR for unequivocal differentiation of all encapsulated and non-encapsulated genotypes of *Trichinella. Int J Parasitol* 29: 1859-1867.

LARVAL TAPEWORM INFECTIONS: CYSTICERCOSIS, CYSTIC ECHINOCOCCOSIS AND ALVEOLAR ECHINOCOCCOSIS

Ana Flisser

Department of Microbiology and Parasitology, Faculty of Medicine, UNAM, Mexico City, Mexico

CYSTICERCOSIS

Etiologic Agent: *Taenia solium* in its larval stage, known as cysticercus.
Clinical Manifestations: neurocysticercosis: a pleomorphic disabling disease of the central nervous system, late onset seizures are the most frequent manifestation. Eye cysticercosis: diminished vision, inflammation, blindness. Muscular and subcutaneous cysticercosis: from asymptomatic with few cysticerci, to muscle pseudo-hypertrophy with massive numbers of parasites.
Complications: minimal to death depending on the number and location of cysticerci and of the degree of inflammation.
Mode of Transmission: Humans acquire cysticercosis after accidentally ingesting *T. solium* eggs released from people that have the adult intestinal worm.
Diagnosis: Imaging techniques, such as magnetic resonance and computed tomography are very sensitive. Although images may not be specific, expertise of the radiologist, will allow identification of brain cysticerci. Immunologic techniques, such as western blot and ELISA, have good sensitivity and specificity, the former being more reliable.
Treatment: praziquantel, albendazole, anti-inflammatory and anti-epileptic drugs, surgery for excision of parasites and for shunt placement.
Reservoir Hosts: Pigs are intermediate hosts and humans are definitive hosts.
Control Measures: Treatment of tapeworm carriers, improvement of sanitation, health education and hygienic pig breeding. Swine vaccination is under study.

Etiology/Natural History

Taenia solium belongs to phylum Platyhelminthes, class Cestoda, subclass Eucestoda, order Cyclophillidea and family Taeniidae. The adult parasite is a flatworm that develops naturally only in humans and measures between 2 and 7 meters long. Tapeworms develop after ingesting cysticerci in raw or undercooked pork meat. The head or scolex evaginates and attaches to the mucosa of the small intestine by its double crown of hooks and its four suckers. The neck, that follows the scolex, continuously forms segments or proglottids that comprise the strobila. Segments may be immature, sexually mature, which are hermaphroditic; and at the end of the strobila, proglottids are gravid and contain around 50,000 eggs each. Eggs are microscopic; they measure around 30 µm and contain a hexacanth embryo surrounded by an oncospheral membrane and an embryophore that is resistant to many environmental conditions. When humans or swine ingest eggs, the gastric and intestinal environment allows desegregation of embryophoric blocks and digestion of oncospheral membrane. Embryos become activated, cross the mucosa of the small intestine, circulate and transform into visible cysticerci in around 3-4 months. Cysticerci, which are the subject of interest of the present chapter, are found mainly in the central nervous system, eye, muscle and subcutaneous tissue. There are two types: cellulose that measures 1-2 cm and is the most common, and the racemose type that measures up to 10 or more cm and is only found in subarachnoidal or ventricular locations of human brain. Cysticerci are comprised of two compartments. The inner compartment contains the scolex and spiral canal, and is surrounded by the outer compartment that contains vesicular fluid, usually less than 0.5 ml (the racemose type may contain up to 90 ml of fluid).

Epidemiology and Control

Cysticercosis is a fascinating disease in its epidemiological component. Most parasitology books show the life cycle, which includes human beings as definitive hosts and pigs as intermediate hosts. Nevertheless, it was always considered that the disease is acquired from eggs ingested in vegetables and fruits irrigated with sewage. Only in the last decade, after several field studies, the main risk factor was identified. That is, the presence of the tapeworm carrier in the household or near-by living facilities. This finding changes the concept of control, since it is more feasible to treat tapeworm carriers than to change sewage management and irrigation infrastructure in developing countries. Cysticercosis is an endemic disease in many countries of Latin America, Sub-Saharan Africa and non-Islamic Asia. Due to immigration there are many patients from developing countries attending hospitals in several USA cities, also *T. solium* carriers have been found in the USA. Therefore cysticercosis is now considered an emerging infectious disease in the USA and transmission of eggs is taking place. As mentioned above, tapeworm carriers are the main element that maintains the life cycle of *Taenia* in the environment and should be treated. However identification of human *T. solium* carriers is remarkably difficult, since the infection is usually asymptomatic. Coproparasitoscopic studies are generally used for diagnosis but lack sensitivity. A coproantigen ELISA with a high sensitivity was developed but is not commercially available, thus other measures should also be pursued. Recent studies suggest that neurocysticercosis may be a risk factor for human cancer. Six percent of cases with malignant hematological disease and 17 percent of patients with glioma had neurocysticercosis. Seemingly, the parasite is capable of immunosuppressing the host's immune response, which in the long term may cause cancer.

Pathogenesis

Pathological aspects of cysticercosis have been described in autopsy and imaging studies performed in humans and pigs. Viable cysts have thin walls; they are filled with clear vesicular fluid and with little or no evidence of surrounding inflammation. It is considered that as parasites loose the ability to control the host immune response, an inflammatory process begins and cysticerci show slight pericystic inflammatory reaction with eosinophils as the first line of attack. Later they become markedly inflamed and edematous, appearing as enhanced ring-like or nodular areas that are granulomatous and in which the vesicular fluid is viscous and opaque. When cysticerci die, they are surrounded by an intense cellular response, which destroys them. Remnants are not detectable by imaging techniques, and they appear as final granulomas in histology sections or become calcified lesions.

Clinical Manifestations and Complications

Cysticerci in brain parenchyma usually generate seizures that can be controlled with adequate antiepileptic drug therapy, excepting occasional cases of parasite growth producing mass effect and compressive manifestations. Extraparenchymal infection may cause hydrocephalus by mechanical obstruction of the ventricles or the basal cisterns by cysticerci or by inflammatory reaction (ependymitis and/or arachnoiditis). Racemose cysticerci follow a progressive course due to their growth. In these cases ventricular shunting is required. Parasites, inflammatory cells and proteins may block shunts. Other complications arise from inflammation, mass effect, or residual scarring and are related to the number and location of cysticerci with an association between surrounding inflammation and development of symptoms, especially in regard to seizures, loss of vision and muscle pseudo-hypertrophy.

Diagnosis

Two types of techniques are used for the diagnosis of neurocysticercosis: imaging and immunologic. Most cities in the USA have imaging equipment, either computed tomography (CT) or magnetic resonance (MR). However this is not the case in developing countries where cysticercosis is one of the main differential diagnoses in patients with neurological symptoms; lack of modern diagnostic support does not hinder adequate diagnosis. In the USA, neurocysticercosis is not usually considered in diagnosis and needs the support of immunologic methods in those cases with images suggestive of neurocysticercosis. Cysticerci may be seen as small non-enhancing roundish images, usually with an invaginated hyperdense or hyperintense scolex, enhancing images, subarachnoid parasites, round calcifications or mixed images with viable and calcified cysticerci, all of them as single, few or multiple images. Western blot is the assay of choice for antibody detection in clinical cases. Between 1 and 7 specific glycoproteins can be identified with the use of serum or cerebrospinal fluid, the former being more sensitive. When more than 2 parenchymal cysticerci are seen by imaging techniques, sensitivity is also higher. Alternatively ELISA is also used, especially in countries where western blot is not available. This technique is less specific and less sensitive, but in cases with neurological symptoms compatible with neurocysticercosis, where there is no echinococcosis, the results are reliable. Antigens of 10, 14 and 18 kDa have been identified and their use for diagnosis is currently being evaluated.

Treatment

Praziquantel (50 mg/kg during 2 weeks) and albendazole (15 mg/kg during 1 week) are effective against *T. solium* cysticerci. Most parenchymal brain cysticerci are killed, but edema and intracranial hypertension may occur, attributed to local inflammation due to death of parasites. In these cases steroids or anticonvulsive drugs are given concomitantly with cestocidal drugs. Care should be taken because serum levels of phenytoin and carbamazepine may decrease during praziquantel administration. Albendazole has better penetration into cerebrospinal fluid and is cheaper; however, recently, one-day treatment with praziquantel has proven to be successful. Imaging studies have shown that some types of parenchymal cysticerci can resolve without being treated with cestocidal drugs, thus it has been suggested that acute, severe brain inflammation caused by their use is unnecessary. On the other hand, allowing a parasite to remain in the brain when there are efficient cestocidal drugs should be questioned, added to the finding in studies performed in pigs that indicate that no calcifications are formed after praziquantel treatment. Surgery is performed to remove cysticerci found in the fourth ventricle and big accessible parasites, and to place ventricular shunts to control hydrocephalus. Dexamethasone, at doses between 4.5 and 12 mg/day is the drug of choice to control inflammation. Prednisone at 1 mg/kg/day may replace dexamethasone when long-term steroid therapy is required. Arachnoiditis or encephalitis can be treated with up to 32 mg/day of dexamethasone. Seizures are usually controlled with any first line antiepileptic drug, which is not withdrawn because of significant risk of relapse in calcified scars. Cysticerci in eyes are usually removed by surgery, while muscle and subcutaneous parasites are eliminated with high efficacy by cestocidal treatment. Care should be taken because multiple disseminated cysticerci may induce uncontrolled anaphylaxis with fatal outcome.

CYSTIC ECHINOCOCCOSIS

Etiologic Agent: *Echinococcus granulosus* in its larval stage, known as hydatid cyst. The disease is now called cystic echinococcosis.
Clinical Manifestations: Liver and lung cysts resemble tumors or abscesses with their concomitant manifestations.
Complications: If cyst ruptures, its contents may cause intense allergic reactions and secondary cysts in other organs.
Mode of Transmission: Humans develop cystic echinococcosis after accidentally ingesting *E. granulosus* eggs released from dogs that harbor the adult intestinal worm.
Diagnosis: Imaging techniques, such as ultrasound, computed tomography and magnetic resonance allow identification of liver and lung cysts. Immunologic techniques, such as western blot, ELISA, agglutination and immunofluorescence have reasonable sensitivity and variable specificity. Commercial kits are based on crude or semi-purified antigens.
Treatment: Surgery to remove parasites, albendazole, mebendazole, PAIR (puncture, aspiration, injection and reaspiration).
Reservoir Hosts: Many mammals, mainly sheep, are natural intermediate hosts and dogs are the definitive hosts.
Control Measures: Vaccination of sheep, control of heard dogs and health education.

Etiology/Natural History

Echinococcus granulosus belongs to phylum Platyhelminthes, class Cestoda, subclass Eucestoda, order Cyclophillidea and family Taeniidae. The adult parasite is a flatworm that develops in dogs and other caniids; it consists of a scolex, a short neck and a mean of 3 proglottids, including one immature, one mature and one gravid segment. The worm measures 2-7 mm long. Tapeworms develop after dogs or foxes ingest hydatid cysts in viscera of the intermediate host. Because each hydatid cyst contains many protoscolices, and each one transforms into a tapeworm, the definitive host harbors thousands of tapeworms. Intermediate hosts become infected, as in cysticercosis, by the ingestion of eggs released in feces. Embryos penetrate the intestinal mucosa and develop into hydatid cysts mainly in the liver and lungs. Cysts are unilocular, sub-spherical, fluid filled and may range in size from a few mm to over 30 cm. Hydatid cysts consist of an inner germinal nucleated layer supported externally by a tough, elastic, acellular laminated layer, surrounded by a host-produced fibrous adventitial layer. Cells bud from the germinal layer into the cystic cavity and produce brood capsules; each one has a diameter of 200-300 μm and contains between 10 and 15 protoscolices. These brood capsules are called hydatid sand; the ones that do not develop protoscolices are sterile. Protoscolices are produced by asexual proliferation. In humans, the slow growing hydatid cyst may attain a volume of many liters and contain many thousands of protoscolices. With time, internal septae and daughter cysts may form within the primary cyst, most probably from protoscolices disrupting the unilocular pattern.

Epidemiology and Control

The greatest prevalence of cystic echinococcosis is in temperate countries, including southern South America, the entire Mediterranean littoral, the Middle East, southern and central parts of the former Soviet Union, central Asia, India, Nepal, China including Tibet, Australia, and parts of Africa. In the USA, most infections are seen in immigrants from countries in which the disease is highly endemic. Sporadic autochthonous transmission is currently recognized in Alaska, Arizona, California, New Mexico and Utah. Geographic strains of *E. granulosus* exist with different host affinities. The northern or sylvatic strain is maintained in wolves and wild cervids (moose and reindeer). Pastoral strains are maintained in dogs and domestic ungulates throughout the world. Populations of *E. granulosus* in different assemblages of hosts (dog-sheep, dog-horse, dog-pig, and dog-cattle) differ in morphology, development, biochemistry, and possibly in infectivity and pathogenicity to humans. The dog-sheep strain, the most widespread variant, is pathogenic in humans. The northern sylvatic strain that occurs in wolf-wild cervid hosts is less pathogenic and the dog-horse cycle, rarely, if ever, infects humans. Probes characterizing the nuclear and mitochondrial DNA of the variant populations provide reliable genetic markers to distinguish them; 9 genotypes have been identified. Globally, sheep are the most important intermediate hosts, but swine, cattle, buffalo, horses and camels are more important in certain regions. Widespread rural practice of feeding viscera of home-butchered sheep to dogs, facilitate transmission of the sheep strain and consequently increase risk of humans to become infected. People acquire cystic echinococcosis through fecal-oral contact, particularly in the course of playful and close contact between children and dogs. Eggs adhere to hair around the infected dog's anus and are also found on muzzle and paws. In endemic areas, preventive measures include personal hygiene, strict dietary regulation of pet dogs to avoid ingestion of sheep offal, and avoidance of street dogs. Periodic prophylactic treatment of pet dogs for intestinal echinococcosis may be useful. Control measures applicable in communities include health education, regulation of livestock slaughtering in abattoirs and farms, control of dogs, and periodic or regular mass treatment of dogs with praziquantel (5 mg/kg) to reduce the prevalence of *E. granulosus* below levels necessary for continued transmission.

Pathogenesis

Cysts evoke an inflammatory reaction of the surrounding tissues that produce the encapsulating fibrous adventitial layer. The impairment of the organs by the common unilocular cyst is chiefly due to pressure. Erosion of blood vessels leads to hemorrhage, and torsion of the omentum leads to vascular constriction. The neighboring tissue cells, depending upon the density of the tissues, undergo atrophy and pressure necrosis as the cyst increases in size. The adventitial layer consists of an inner zone of epithelioid cells (epithelioid zone), and an outer one of granulation tissue (fibrous zone). Fibroblasts are loosely arranged and eosinophils, neutrophils, lymphocytes and histiocytes infiltrate the fibrous zone. Cyst fluid may be clear, yellowish, brown, and colloidal or even a brown caseous material depending on the degenerating process of cysts. Death of cysts results in disintegrating parasite tissue including hooks in the cyst, calcification, and an intense cellular infiltrate, mainly of eosinophils. During the natural course of infection, the fate of *E. granulosus* cysts is variable. Cysts may grow to a certain size and then persist without noticeable changes for many years or even some decades. Other cysts may spontaneously rupture to collapse and disappear.

Clinical Manifestations and Complications

Symptomatology in cystic echinococcosis is related to parasite biomass. Signs and symptoms of hepatic echinococcosis include hepatomegaly with or without a palpable mass, liver abscesses, calcified lesions, epigastric pain, nausea, and vomiting, portal hypertension, inferior vena cava compression or thrombosis, secondary biliary cirrhosis, biliary colic-like symptoms, biliary peritonitis or fistula formation. If a cyst ruptures, the sudden release of its contents may precipitate allergic reactions ranging from mild to fatal anaphylaxis, bacterial infection may occur, and since there is spread of protoscolices, a multiple secondary echinococcosis disease may start. Clinical symptoms of pulmonary cystic echinococcosis include tumor-like chest pain, chronic cough, fever, haemoptysis, pneumothorax, pleuritis, lung abscess, eosinophilic pneumonitis, lung embolism and sometimes expectoration. Because of the slowly growing nature of cysts, most cases of liver and lung cysts as well as those found in heart, spine and brain, and even bone, are diagnosed in adult patients and symptoms are related to tumor-like lesions.

Diagnosis

The presence of a cyst-like mass in a person with a history of exposure to sheepdogs in areas in which *E. granulosus* is endemic supports the diagnosis of cystic echinococcosis. Ultrasonography (USG), CT and MR are the techniques of choice. Images typically show round, solitary or multiple, sharply contoured cysts, measuring from 1 to over 15 cm, the presence of internal daughter cysts produces structures comparable to a cartwheel or as undulated membranes floating within the cyst fluid. Thin, crescent or ring-shaped calcifications in the cyst can also be seen; these require about 5-10 years to develop. Cysts have been organized into five types based on USG findings: simple univesicular or hyaline hydatid cysts are type 1; they occur preferentially in asymptomatic persons and are considered an early developmental stage. Cysts with a clear laminated membrane and/or presence of daughter cysts correspond to types 2 and 3, while hydatid cysts showing signs of involution such as infiltration and calcification are included in types 4 and 5. Computed tomography and MR allow measurement of the size of the parasites, which is useful for chemotherapeutic follow-up and gives a correct diagnosis in a high proportion of cases. Serology is useful for confirmation of radiographic findings. Hepatic cysts are more likely to elicit an immune response than pulmonary cysts, but about 10 percent of patients with hepatic cysts and 40 percent with pulmonary cysts do not produce detectable serum antibodies and give false negative results, which may result in a dangerous puncture of a hydatid cyst. Most routine laboratory test systems or commercial kits are based on crude or semi-purified antigens. These are reasonably sensitive but specificity is not always satisfactory. The use of the two major hydatid cyst fluid antigens, antigen 5 (thermolabile) and antigen B (thermostable), is restricted to scientific applications. Both are lipoproteins composed of subunits. In antigen 5, subunits of 52-67 kDa were identified under non-reducing conditions, while subunits of 20-24 and 38 kDa were detected under reducing conditions. Antigen B, purified from human hydatid cyst fluid by the method of Oriol exhibited a high sensitivity and specificity in ELISA. It is considered that antigen B is more specific than antigen 5 for diagnosis of *E. granulosus*. Similar sensitivity and specificity have been found with native and recombinant antigen B.

Treatment

Surgical removal of hydatid cysts is performed in most patients with few complications and good prognosis, even though it is impossible to know whether an operation has removed all parasite tissue. Surgery is the preferred treatment when cysts are large or secondarily infected. The aim of surgery is total removal of the cyst, carefully avoiding the adverse consequences of spilling its contents. Pericystectomy is the usual procedure but removal of a section or the whole organ involved may be used, depending on the location and condition of the cyst. Chemotherapy may be routinely carried out for at least 2 years after radical surgery, with careful monitoring of the patient during a minimum of 10 years for possible recurrence. Albendazole (10 mg/kg) and mebendazole (40-50 mg/kg) are effective; however, because of a better intestinal absorption and penetration of albendazole into the cysts, it is slightly more efficient. Chemotherapy in children is more successful than in adults. Adverse reactions, such as neutropaenia, proteinuria, mild hepatotoxicity, gastrointestinal disturbances and transient alopecia, are reversible upon cessation of treatment. A minimum period of treatment is 3 months, but long-term prognosis in individual patients is difficult to predict, therefore, prolonged follow up with USG or other imaging procedures is needed to determine the eventual outcome. Types 1 and 2 cysts respond best to cestocidal drugs, whereas those grouped in types 3-5 are relatively refractory to treatment. Difficult presentations, such as bone cysts, may also respond favorably to chemotherapy. Puncture, aspiration, injection and reaspiration (PAIR) is highly recommended. Percutaneous puncture is performed under ultrasonographic guidance, followed by aspiration of the liquid contents, instillation of a protoscolicidal agent (ethanol or hypertonic sodium chloride) and drainage. The procedure and can be done by laparoscopy. To avoid sclerosing cholangitis, this procedure is performed only in patients whose cysts do not have biliary communication. The possibility of secondary echinococcosis resulting from accidental spillage during surgery or PAIR can be minimized by concurrent treatment with albendazole and by carefully washing the abdominal cavity with the cestocidal solution.

ALVEOLAR ECHINOCOCCOSIS

Etiologic Agent: *Echinococcus multilocularis* in its larval stage, known as alveolar hydatid cyst. The disease is now called alveolar echinococcosis.
Clinical Manifestations: Cysts develop mainly in the liver causing cirrhosis or mimicking a carcinoma. Due to their capacity for tumor-like proliferation and potential for metastasis, development can also occur in different organs with concomitant manifestations.
Complications: Alveolar echinococcosis resembles malignant tumors in morphology and behavior, thus complications are diverse and important, and mortality is high.
Mode of Transmission: Humans develop alveolar echinococcosis after accidentally ingesting *E. multilocularis* eggs released from cats or foxes that harbor the adult intestinal worm.
Diagnosis: Imaging techniques allow detecting cysts in practically any location, immunology assays such as western blot and ELISA are performed with recombinant EM10 antigen or native Em18 antigens.
Treatment: Surgery to remove parasites, albendazole and mebendazole.
Reservoir Hosts: Rodents are natural intermediate hosts and foxes are the definitive hosts. Other carnivores may also be definitive hosts.
Control Measures: Control of heard dogs and foxes used for hunting, and health education.

Etiology/Natural History

Echinococcus multilocularis belongs to phylum Platyhelminthes, class Cestoda, subclass Eucestoda, order Cyclophillidea and family Taeniidae. The adult parasite is a flatworm that develops mainly in foxes. It consists of a scolex, a short neck and a mean of 4-5 proglottids, including immature, mature and one gravid segment. The worm measures 1.2 to 4.5 mm long. Tapeworms develop after ingesting hydatid cysts in viscera of intermediate hosts. Because each alveolar hydatid cyst contains many protoscolices, and each one transforms into a tapeworm, the definitive host harbors thousands of tapeworms. Intermediate hosts become infected by the ingestion of eggs released in feces. Embryos penetrate the intestinal mucosa and develop into alveolar hydatid cysts mainly in liver. *Echinococcus multilocularis* is the most complex of larval cestodes and develops quite differently from that of *E. granulosus*. The size of each parasite varies from less than 1 mm to 20 mm, but the lesions caused vary from minor foci (a few millimeters in diameter) up to large areas of infiltration. The parasite is a multi-vesicular structure with no limiting host-parasite barrier (without adventitial layer). The larval mass usually contains a semisolid matrix rather than fluid and is formed by numerous small vesicles embedded in a dense stroma of connective tissue, consisting of a network of filamentous solid cellular protrusions of the germinal layer that are responsible for infiltrating growth, transforming into tube-like and cystic proliferating structures. In rodents, the natural intermediate hosts, the larval mass proliferates rapidly by budding tissue and produces an alveolar-like pattern of microvesicles filled with protoscolices. In humans the larval mass resembles a malignant tumor because it also invades surrounding tissues. Protoscolices are rarely observed.

Epidemiology and Control

The life cycle of *E. multilocularis* involves foxes (mainly red fox and arctic fox) and their rodent prey, such as voles, lemmings, hamsters, gerbils, rats, mice, shrews, muskrats and pikas, in ecosystems generally separated from humans. Thus, exposure of humans to *E. multilocularis* is relatively less common than exposure to *E. granulosus*. There is, nevertheless, an ecological overlap with humans, because domestic dogs or cats may become infected when they eat infected wild rodents. Alveolar hydatid disease has been reported in parts of central Europe, in eastern Russia, the Central Asian republics, western China, Japan, the north western portion of Canada and western Alaska. Wildlife transmission, involving red foxes, coyotes and prairie voles, appears to be increasing in central USA. Hunters, trappers, and persons who work with fox fur may be exposed to alveolar echinococcosis but the main risk factors appear to be farming or agricultural occupations and a history of dog ownership. Hyperendemic foci have been described in some Eskimo villages of the North American tundra and in China where local dogs regularly feed on infected commensal rodents. The epidemiology of this disease in China remains one of the major challenges in temperate areas. Eliminating *E. multilocularis* from its wild animal hosts is impossible; therefore, contact with dogs and foxes in areas where the infection is endemic should be avoided. Infections in dogs and cats prone to eat infected rodents can be prevented by monthly treatment with praziquantel. Preventing infections in humans depends on education to improve hygiene and sanitation. Alveolar echinococcosis is a disease associated with advanced age, with a peak at 60-70 years as compared to 30-40 in cystic echinococcosis.

Pathogenesis

An inner necrotic zone and outer layers of histiocytes and lymphocytes surround cysts. In later stages, tissue reactions of chronic inflammation, often with giant-cell foreign body reaction, fibrous tissue or necrotic areas are seen around cysts. Fibrous proliferation is often so intense that cysts are embedded in a very dense and hard fibrous stroma. However, the parasite as a whole is not demarcated at its outer limits by a fibrous capsule like the cysts of *E. granulosus*. Advanced lesions consist of a central necrotic cavity filled with a white amorphous material that is covered with a peripheral layer of dense fibrous tissue. There are focal areas of calcification and extensive infiltration by proliferating vesicles. Brood capsules and protoscolices are rarely formed, which suggests that the human host does not provide optimal conditions for the development of the parasite. In many cysts a germinal layer is not discernible in light microscopy or it appears as a thin and delicate layer with only a few nuclei. Nevertheless, as shown by transplantation to animals, such parasites still have potential for proliferation.

Clinical Manifestations and Complications

The primary location of alveolar echinococcosis is the liver but the capacity for tumor-like proliferation and the potential for metastasis means that development can also occur in lung, brain, bone or other organs. Infection of the liver closely mimics hepatic carcinoma or cirrhosis. Clinical cases are characterized by a long lasting chronic course of the disease. The initial symptoms of alveolar echinococcosis are usually vague. Mild upper quadrant and epigastric pain with hepatomegaly may progress to obstructive jaundice. Occasionally, initial manifestations are related to metastases to the lungs or brain. Mortality may be high and rises to 100% in untreated or inadequately treated patients. Early-stage disease can be detected by USG or serologic screening of people with high-risk of acquiring alveolar echinococcosis. Early detection also improves prognosis through earlier application of therapy.

Diagnosis

As is the case for cysticercosis and cystic echinococcosis, diagnosis of alveolar echinococcosis is generally performed by imaging techniques. The usual CT image of *E. multilocularis* infection is that of solid tumors seen as heterogeneous hypodense masses often associated with central necrotic areas and calcifications. Lesion contours are irregular without a well-defined wall. Frequently the lesion extends beyond the liver, which can cause compression or obstruction of other structures such as the inferior vena cava, hepatic veins and portal branches. Microcalcifications can be aggregated or can be seen as irregular plaque-like calcified foci in central or peripheral parts of the lesions. Lungs may be involved either by direct extension of the liver process or by parasite metastases, seen as multiple small solid foci located usually eccentrically at the periphery of the lobes. Serologic tests are positive at high titers; most patients exhibit antibodies against the recombinant EM10 antigen, a 65-kDa molecule. Also an 18-kDa native antigen, Em18, is highly species specific and sensitive in immunodiagnosis, as shown by western blot.

Treatment

Surgical resection of the involved liver segment and of parasite lesions from other affected organs is indicated in all operable cases, although it is impossible to predict whether an operation has removed all the parasite. Therefore postoperative chemotherapy is now a common practice for at least 2 years after radical surgery with careful monitoring of the patient during a minimum of 10 years for possible recurrence. Long-term treatment with mebendazole (50 mg/kg) or albendazole (10 mg/kg) inhibits growth of larval *E. multilocularis*, reduces metastasis, and enhances quality and length of survival. Liver transplantation has been employed in some otherwise terminal cases with good survival rate, although immunosuppression following transplantation may enhance proliferation of hydatid cysts in other organs.

RECENT ADVANCES AND CONTEMPORARY STUDIES

The data presented in Table 1 allow identifying trends in contemporary studies and show recent advances. As expected, clinical studies that inform about cases with common and uncommon localizations of cysticerci and hydatid cysts account for 37% of all articles published in 2000 and 2001. Please note that since this type of report does not allow distinguishing between cystic and alveolar echinococcosis, most references were included in the former and those that specifically state alveolar echinococcosis are listed in the last column. Interestingly there is a similar number of publications on diagnosis; but since there are two times more articles on echinococcosis than on cysticercosis, this suggests that relatively more research is taking place on diagnosis of cysticercosis. This research is mostly on immunodiagnosis of patients, including evaluation of purified antigens, recombinant proteins and synthetic peptides, as well as on detection of definitive hosts; descriptions of imaging findings are also included. On the contrary, articles on treatment are more abundant on cystic and alveolar echinococcosis. This reflects the lack of optimal surgical and especially cestocidal treatment for both types of echinococcosis. Epidemiology and control studies make almost 40% of all publications that specify alveolar echinococcosis and are related mainly to foxes and other hosts. The areas related to basic studies include publications on immunology, which refer mainly to vaccination of swine and a few to characterization of the immune response. Several articles were on biology of these tapeworms and application of molecular biology techniques to their study. Interestingly 17 general review articles were published in 2 years.

Table 1. Number of Articles Published In 2000 and 2001 by Disease and Area of Research

Area of Research	Cysticercosis	Cystic Echinococcosis	Alveolar Echinococcosis
Biology	10	22	4
Molecular biology	6	18	4
Immunology	31	27	6
Diagnosis	44	58	7
Treatment	18	99	13
Clinical aspects	88	190	5
Epidemiol & control	25	33	25
General reviews	5	10	2
TOTAL	222	447	64

REFERENCES

Books

Aluja A, Escobar A, Escobedo F, Flisser A, Laclette JP, Larralde C, Madrazo I, Velazquez V and Willms K. 1987. *Cisticercosis: Una recopilación actualizada de los conocimientos básicos para el manejo y control de la cisticercosis causada por Taenia solium.* México, DF, México: Fondo de Cultura Económica. Serie Biblioteca de la Salud. México. 115 pp.

Andersen FL, Chai, JJ, Liu FJ (Eds). *Compendium on Cystic Echinococcosis with Special Reference to the Xinjian Uygur Autonomous Region, The People's Republic of China.* Provo, Utah: Brigham Young University. 235 pp,

Andersen FI, Ouhelli H and Kachani M. 1997. *Compendium on Cystic Echinococcosis in Africa and in the Middle Eastern Countries with Special Reference to Morocco.* Provo, Utah: Brigham Young University.
Arriagada C, Nogales-Gaete J, Apt W (Eds). 1997. *Neurocisticercosis.* Arrynpog Ediciones, Santiago de Chile, Chile: Arrynpog Ediciones. 333 pp.
Craig P, Pawlowski Z (Eds). 2002. *Cestode Zoonoses: Echinococcosis and Cysticercosi. An Emergent and Global Problem.* NATO Science Series. Series 1: Life and Behavioural Sciences Vol. 341. Amsterdam, The Netherlands: 395 pp.
Del Brutto OH, Sotelo J, Roman G. 1998. *Neurocysticercosis: a clinical handbook.* The Netherlands: Lisser, Swets & Zeitlinger. 207 pp.
Eckert J, Gemmell MA, Meslin FX, Pawlowski ZS. 2001. *WHO/OIE Manual on Echinococcosis in Humans and Animals: a Public Health Problem of Global Concern.* Paris , France: World Organization for Animal Health. 265 pp.
Eckert J, Gemmell MA, Matyas Z and Soulsby EJL (Eds). 1994. *Guidelines for Surveillance, Prevention and Control of Echinococcosis/hydatidosis,* 2nd ed. Geneva, Switzerland: World Health Organization.
Flisser A, Madrazo I, Delgado H. 1997. *Cisticercosis Humana.* Mexico, DF, México: El Manual Moderno. 176 pp.
Flisser A, Malagon F (Eds). 1989. *Cisticercosis Humana y Porcina. Su Conocimiento e Investigación en Mexico.* Mexico DF: Editorial Limusa Noriega. 266 pp.
Flisser A, Van der Kaay HJ, Van Knapen F, Overbosh D (Eds). 1989. *C-Now. Proceedings of the Symposium on Neurocysticercosis.* Rotterdam, Leiden, The Netherlands: Acta Leidensia, Association Institute of Tropical Medicine Vol. 57. 273 pp,
Flisser A, Willms K, Laclette JP, et al. (Eds). 1982. *Cysticercosis. Present State of Knowledge and Perspectives.* New York: Academic Press. 700 pp.
Garcia HH, Martínez SM (Eds). 1999. *Taenia solium Taeniasis/Cysticercosis.* Lima Peru: Editorial Universo. 346 pp.
Gemmell M, Matyas Z, Pawlowski Z, Larralde C. 1983. *Guidelines for Surveillance, Prevention and Control of Taeniasis/Cysticercosis.* VPH/83.49. Geneva, Switzerland: World Health Organization. 207 pp.
Palacios E, Rodriguez-Carbajal J, Taveras JM (Eds). 1983. *Cysticercosis of the Central Nervous System.* Springfield, IL: Thomas.
Ruiz A, Schantz P, Arambulo P III (Eds). 1995. *Proceedings of the Scientific Working Group on the Advances in the Prevention, Control and Treatment of Hydatidosis.* Washington, D. C.: Pan American Health Organization.
San Esteban JE, Flisser A., Gonzalez-Astiazaran A (Eds). 1997. *Neurocisticercosis en la infancia.* Porrua, México DF: Grupo Editorial MA. 317 pp.
Thompson RCA, Lymbery AJ (Eds). 1995. *Echinococcus and Hydatid Disease.* Wallingford, UK: CAB International. 477 pp.
Uchino J, Sato N (Eds). 1993. *Alveolar Echinococcosis of the Liver.* Sapporo: Hokkaido University School of Medicine. 215 pp.

Book chapters

Escobedo F, Gomez-Aviña A, Ruiz-Gonzalez S. 2000. Neurosurgical aspects of Neurocysticercosis. In: Schmidek HH (Ed.), *Operative Neurosurgical Techniques,* 4th ed. Philadelphia, Pennsylvania: WB Saunders Co. pp. 1756-1768.
Flisser, A. 1998. Larval Cestodes. In: Cox FEG, Kreier JP, Wakelin D (eds.), *Microbiology and Microbial Infections,* Vol. 5, 9th ed. Topley & Wilson's. pp 539-560. (A second edition of this chapter can be found in Topley Online, 2001 by Flisser A and Craig P).
Flisser A. 1995. *Taenia solium, Taenia saginata* and *Hymenolepis nana.* In: MJG Farthing, GT Keusch, D Wakelin (Eds.), *Enteric Infection 2. Intestinal Helminths.* London, UK: Chapman & Hall Medical. Pp. 173-189.

Flisser A. 1994. Taeniasis and Cysticercosis due to *Taenia solium*. In: T. Sun (Ed.), *Progress in Clinical Parasitology, Vol. 4*. Boca Raton, Florida: CRC Press.pp. 77-116.

Flisser A, Larralde C. 1986. Cysticercosis. In Walls KW, Schantz PM (Eds.), *Immunodiagnosis of Parasitic Diseases*. Orlando, Florida: Academic Press. pp. 109-161.

Gemmell MA, Roberts MG. Cystic echinococcosis (*Echinococcus granulosus* In: Palmer SR, Soulsby L, Simpson IH (Eds), *Zoonoses*. Oxford UK: Oxford University Press. pp. 665-688.

Grove DI. 1990. *Echinococcus granulosus* and Echinococcosis or Hydatid Disease. In *A History of Human Helminthology*. Wallingford, Oxon, UK: CAB International. pp 319-353.

Grove DI. 1990. *Taenia solium* and taeniosis and cysticercosis. In *A History of Human Helminthology*. Wallingford, Oxon, UK: CAB International. pp. 355-383

Houin R, Flisser A, Liance AM. 1994. Cestodes larvaires. In: *Maladies Infectieuses*. Encyclopédie Médico-Chirurgicale. París, France.. 8-511-A-10, 1-22.

Lightowlers MW, Mitchell GF, Rickard MD. 1993. Cestodes In: Warren KS, Agabian N (Eds.), *Immunology and Molecular Biology of Parasitic Infections*, 3rd ed. Boston, MA: Blackwell Scientific. Pp. 438-72.

Macpherson CNL, Craig PS. 2000. Dogs and cestode zoonoses. In: CNL Macpherson, FX Meslin, AI Wandeler (Eds.), *Dogs, Zoonoses and Public Health* Wallingford, Oxon, UK: CABI Publishing. pp 177-211.

Madrazo I, Flisser A. 1992. Parasitic infestations of the cerebrum. Cysticercosis In: Appuzo JML (Ed), *Brain Surgery. Complication Avoidance and Management*. Edinburgh, UK: Churchill Livingston. pp. 1419-1430.

Mitchel GF. 1990. Vaccines and vaccination strategies against helminthes. In: Agabian N, Cerami A (Eds), *Parasites. Molecular Biology, Drug and Vaccine Design*. New York: Wiley-Liss. pp. 349-63.

Schantz PM. 1994. Larval cestodiases. In: Hoeprich PD, Jordan MC, Ronald AR (Eds.), *Infectious Diseases. A Treatise of Infectious Processes*, 4th ed. Philadelphia, Pennsylvania: JB Lippincott. pp 850-860.

Schantz PM, Gottstein B. 1986. Echinococcosis (hydatidosis). In: Walls KW, Schantz PM (Eds.), *Immunodiagnosis of Parasitic Diseases, Vol. 1. Helminthic Diseases*. Orlando, FL: Academic Press. pp 69-107.

Schantz PM, Okelo GBA. 1990. Echinococcosis (Hydatidosis). In: Warren KS, Mahmoud AAF (Eds.), *Tropical and Geographical Medicine*, 2nd ed. Cleveland, OH: University Hospitals of Cleveland. pp 504-518.

Schantz PM, Wilkins PP, Tsang VCW. 1998. Immigrants, imaging and immunoblots: the emergence of neurocysticercosis as a significant public health problem. In: Scheld WM, Craig WA, Hughes JM (Eds), *Emerging Infections 2*. Washington DC: ASM Press. pp. 213-242.

White C, Robinson P, Kuhn R. 1997. *Taenia solium* Cysticercosis: Host-Parasite Interactions and the Immune Response. In: Freedman DO (Ed.), *Immunopathogenic Aspects of Diseases Induced by Helminth Parasites, Chemical Immunology*. Basel, Switzerland: Karger. pp. 209-230.

Williams K. 2001. Cestodes. In: . Gillespie S, Pearson RD (Eds.), *Principles and Practice of Clinical Parasitology*. New York: John Wiley & Sons Ltc. pp. 613-633.

INTESTINAL TAPEWORM INFECTIONS

Dennis J. Richardson

Quinnipiac University, Hamden, Connecticut

Etiologic Agents: *Hymenolepis nana*, *Hymenolepis diminuta*, *Dipylidium caninum*, *Mesocestoides* spp., and *Diphyllobothrium* spp.
Clinical Manifestations: Most intestinal tapeworm infections are light and therefore are asymptomatic. Symptomatic infection is typically mild and ill-defined. The most common symptoms include diarrhea, abdominal discomfort, anorexia, and nausea.
Complications: Potential hyperinfection with *H. nana* due to autoinfection is possible, especially in immunocompromised patients.
Mode of Transmission: Ingestion of cysticercoids contained in intermediate hosts, which include grain beetles for *Hymenolepis* spp. Ingestion of eggs in food or water contaminated by rodent feces also may result in transmission of *H. nana*. Fleas serve as intermediate host for *D. caninum*. The life cycle, and thus the intermediate host(s) has not been determined for any species in the genus *Mesocestoides*, although various species of rodents and reptiles may serve as paratenic (transport) hosts, the ingestion of which may lead to human infection. Infection with *Diphyllobothrium* sp. occurs through the ingestion of the larval plerocercoid in raw or poorly cooked fish.
Diagnosis: Observation of eggs or proglottids in stool
Drugs of Choice: Praziquantel or niclosamide
Reservoir Hosts: Rodents for *Hymenolepis* spp., dogs and cats for *D. caninum*, various mammals for *Mesocestoides* spp., and various fish eating mammals for *Diphyllobothrium* spp.
Control Measures: Avoid eating infected intermediate hosts, including grain beetles and fleas. Treat infected pets and control flea infestation to control *D. caninum*. Cook fish thoroughly to avoid infection with *Diphyllobothrium* spp.

ETIOLOGY/NATURAL HISTORY

The cestode order Cyclophyllidea contains the majority of tapeworm species for which humans serve as the definitive host. Within the small intestine, worms attach, mature, mate and produce eggs that are passed in the feces. After being ingested by the intermediate host, eggs hatch in the small intestine and release a hexacanth larva. The larva penetrates the gut wall and gains entry into the hemocoel, where it develops into a cysticercoid (the infective stage to the definitive host). Transmission occurs when a potential definitive host ingests cysticercoids contained within an intermediate host. Grain beetles (order Tenebrionidae) and fleas typically serve as intermediate hosts for *Hymenolepis* spp. and *D. caninum*, respectively. Although chewing lice may also serve as intermediate hosts for *D. caninum*, fleas (especially *Ctenocephalides felis* and *Ctenocephalides canis*) are of far greater epidemiological importance.

***Hymenolepis* spp.**

Rodents serve as the normal definitive hosts for *Hymenolepis* spp. Traditionally, mice have been considered to be the "primary" host for *Hymenolepis nana* (syn. *Vampirelopis nana*) and rats were considered to be the primary host for *H. diminuta.* In a survey of pet store rodents, however, Duclos and Richardson (2000) found that rats were more frequently utilized as definitive hosts for *H. nana* than were mice. Grain beetles (family Tenebrionidae) are the most common intermediate host for *Hymenolepis* spp., although fleas are also competent intermediate hosts. Infective cysticercoids may be found in beetles and fleas by 3 weeks post-infection. In addition to serving as definitive host for *H. nana*, rodents and humans may also become infected by ingesting *H. nana* eggs. In the latter route, eggs hatch in the intestine and release the hexacanth, which burrows into the intestinal mucosa. Within the mucosa, the oncosphere transforms into the cysticercoid, which then emerges into the small intestine in 5 to 6 days and develops into the adult. From the time that the cysticercoid arrives in the lumen of the intestine, by either route, the prepatent period in the definitive host is about 3 weeks.

Autoinfection takes place when eggs hatch before they are expelled from the colon. Autoinfection occurs in laboratory mice and may possibly occur in humans. The hexacanths liberated from the eggs enter the mucosa of the intestine and develop into cysticercoids. Such completion of the life cycle within the host may lead to massive infections. In immunocompetent hosts, the process of autoinfection is kept in check by the immune response directed against the hexacanths and cysticercoids within the host tissue (Heyneman, 1962; Isaak et al., 1977). This immune response prohibits extremely large "superinfections." When the host is immunosuppressed , however, massive infections may develop and metastasis of cysticercoids to other organs may lead to disseminated infection. This is potentially a very serious condition, similar to that of disseminated strongylodiasis. Such "superinfections" with dissemination of cysticercoids has been demonstrated in immunosuppressed mice (Okamato, 1969; Lucas et al., 1980). Disseminated hymenolepiasis has not been clearly documented in humans, but Lucas et al. (1979) suggested that *H. nana* cysticercoids might have been responsible for a disseminated parasitic infection in a Pennsylvania man with Hodgkin's disease who had been treated with immunosuppressive drugs and radiotherapy. Until the etiology of hymenolepiasis in the immunocompromised human is better understood, infection with *H. nana* should be considered a serious threat to the immunocompromised patient.

It has been suggested (Al-Baldawi et al., 1989) that different strains of *H. nana* exist in rodents and humans and that the two are highly host specific. Data supporting this claim are conflicting and inconclusive and this tapeworm is typically classified as a zoonotic parasite, fully capable of horizontal transmission between humans and non-human animals (Duclos and Richardson, 2000).

Dipylidium caninum

The normal hosts for *D. caninum* are dogs and cats. Capsules with 8-15 eggs each are contained within proglottids, and the proglottids are passed in feces. *Dipylidium caninum* is a common tapeworm of both dogs and cats worldwide. Prevalence rates throughout the United States and Canada range from 1 to32 percent in cats (Rembiesa and Richardson, 2003). This estimate is likely to be lower than the actual prevalence because the most common technique used in helminth surveys of cats and dogs is fecal floatation which is highly inefficient in the detection of *D. caninum* infection (Blagburn, 2001). The best means of detecting *D. caninum* infection is the detection of proglottids in the stool. No good estimates are available concerning the prevalence of *D. caninum* in North American dogs, but the consensus is that it is high. Egg packets are expelled from the proglottid in the stool as the proglottid begins to dry out. Larval fleas ingest the eggs while feeding on organic debris, including feces. The larval tapeworms survive the metamorphosis of the larval fleas to adults. Dogs, cats, and humans become infected by ingesting infected fleas. The prepatency period in the definitive host is about 25 days (Despommier et al., 2000).

***Mesocestoides* spp.**

Several species in the genus *Mesocestoides* have been described from a broad array of mammalian and avian hosts. *Mesocestoides* spp. are the most common tapeworms of medium-sized furbearing mammals such as raccoons, skunks and opossums. Various amphibians, reptiles, birds, and mammals may serve as second intermediate or paratenic hosts. No complete life cycle has been described for any member of the genus, but it is assumed that an arthropod intermediate host is required. The mode of transmission to humans is unknown.

***Diphyllobothrium* spp.**

Diphyllobothrium spp. belong to the cestode order Pseudophyllidea. Pseudophyllidean tapeworms exhibit an indirect life cycle that requires two intermediate hosts for completion. Tapeworms attach, mature, mate, and produce eggs in the small intestine of the definitive host. Eggs are released singly through the genital pore of the tapeworm and are then passed in the host's feces. In water, eggs hatch and release a free-swimming ciliated coracidium that may be ingested by the first intermediate host, a copepod (a minute crustacean). In the copepod, the oncosphere travels from within the coracidium through the gut wall into the hemocoel where it develops into a procercoid. Fishes that serve as second intermediate hosts become infected when they ingest the procercoid contained within the copepod. The plerocercoid develops in muscle tissue and a definitive host becomes infected when it ingests a plerocercoid in raw or poorly cooked fish. The adult tapeworm develops rapidly within the definitive host. After eggs have been passed from the proglottids, strands of "senile" proglottids break off and are passed in the feces.

Several species of *Diphyllobothrium* occur in the Americas. It is impossible to distinguish between species based on morphology of eggs or proglottids. Virtually any piscivorous (fish eating) mammal may serve as a definitive host for these tapeworms. Some species (such as *D. latum*) are involved in a freshwater cycle of transmission while others (such as *D. pacificum*) utilize a marine cycle.

EPIDEMIOLOGY

An analysis of data from 216,275 stool specimens examined by state diagnostic laboratories throughout the United States conducted by Kappus et al. (1991) provides a good comparative analysis of the prevalence of patent zoonotic tapeworm infections occurring in the United States. *Hymenolepis nana* was by far the most common tapeworm occurring in the United States. It was observed in 900 samples for a nation-wide prevalence of 0.4 percent. *Hymenolepis diminuta* was observed in 16 samples with a prevalence of less than 1/100th of 1 percent. *Diphyllobothrium latum* and *D. caninum*, observed in 14 and 11 stool samples, respectively, were equally as rare. The vast majority of intestinal cestodiasis occur in children, probably because they engage in behavior that facilitates transmission. There are no apparent physiological or immunological factors prohibiting infections in adults.

Hymenolepis spp.

Hymenolepis nana is one of the most common tapeworms of humans, infecting an estimated 75 million individuals worldwide (Crompton, 1999). In India, 50 percent of children have been reported to be infected (Marty and Neafie, 2000a). *Hymenolepis nana* is relatively common in Mexico (Flores, 1983; Guarner et al., 1997) and is the most common tapeworm of humans in the United States. In the 1987 survey by Kappus et al. (1991), *Hymenolepis nana* was reported from 34 states, with local prevalences as high as 1.6 percent (Rhode Island). It is likely that direct transmission (facilitated by ingestion of the egg) is responsible for most human cases (Turner, 1975; Duclos and Richardson, 2000).

Relatively little is known about the prevalence of *H. nana* "in nature." People contract *H. nana* by ingesting infected grain beetles, or by ingesting eggs from grain stores or products contaminated by rodent feces. The latter seems less likely because the eggs do not remain viable for long in feces, once outside the host (Maki and Yanagisawa, 1987). Humans may also become infected by direct transmission of eggs on the fur of pet rodents or in food or water contaminated with feces from infected humans or rodents. Transmission by oral-anal contact is also possible. Pet rodents, especially rats, have been implicated as potentially important sources of zoonotic transmission to humans. Duclos and Richardson (2000) found that 75 percent of pet stores in south-central Connecticut were selling rodents, particularly rats, which were infected with *H. nana*. Overall, rats, mice and hamsters in surveyed pet stores exhibited prevalences for *H. nana* of 32 percent, 16 percent, and 10 percent respectively. *Hymenolepis diminuta* was found in a single rat for a prevalence of 2.6 percent.

Hymenolepis nana poses a particular threat to immunocompromised persons because of the potential development of massive infections through autoinfection and subsequent disseminated hymenolepiasis. Conti et al. (1995) reported that 2 percent of Florida AIDS patients owned pet rodents, but in most instances health care workers had failed to advise them about zoonotic threats associated with companion animals. Guarner et al. (1997) found that 9 percent of oncologic patients receiving chemotherapy at the National Cancer Institute in Mexico City were infected with *H. nana*. It is crucial that any patient with an immune deficiency, or that is undergoing immunosuppressive therapy, be counseled about preventing parasitic infection.

Hymenolepis diminuta rarely infects humans (Hamrick et al., 1990). The difference in prevalence between human infections with *H. nana* and *H. diminuta* is probably indicative of the importance of direct infection of humans by eggs of *H. nana*. *Hymenolepis diminuta* requires a coleopteran intermediate host for completion of the life cycle. The only way that humans may become infected is by ingestion of the cysticercoid in an infected beetle.

Dipylidium caninum

Humans become infected with *D. caninum* by ingesting infected fleas containing cysticercoids. Although some chewing lice have demonstrated the ability to serve as competent intermediate hosts for *D. caninum*, they are probably epidemiologically insignificant. The most common fleas involved are the dog flea (*C. canis*) and the cat flea (*C. felis*) with the latter being more important as it is the most common flea infesting both dogs and cats.

Human infection is rare and almost always occurs in children, although it has been suggested that human infections may be underdiagnosed and misdiagnosed due to a lack of awareness of human dipylidiasis on the part of physicians (Turner, 1962; Chappell et al., 1990). Infection with *D. caninum* in pets or humans is indicative of flea infestation; thus, in addition to treatment of the helminth infection, the flea infestation should also be addressed.

***Mesocestoides* spp.**

Only six cases of human mesocestodiasis have been reported in North America from Texas, Missouri, New Jersey, Ohio, Mississippi, and California (Chandler, 1942; Gleason and Healy, 1967; Gleason et al., 1973; Gutierrez et al., 1978; Hutchinson and Martin, 1980; Schultz et al., 1992). Until the taxonomy and especially the life cycle of these tapeworms is resolved, epidemiological factors leading to human infection will remain a mystery.

***Diphyllobothrium* spp.**

Although several species of *Diphyllobothrium* occur in the Americas, the species of *Diphyllobothrium* that is best known in the United States is *D. latum*. Desowitz (1981) provided a very colorful account of how this worm was imported into the Great Lakes region by immigrant fishermen from Scandinavian countries during the 19th century and of "Jewish grandmothers" acquiring infections with this parasite by sampling gefilte fish prior to cooking. Of 216,275 stool samples examined in 1987, only 14 were positive for *D. latum*. Infections were reported from Wisconsin (3), Minnesota (2), North Carolina (1), Texas (2), Oregon (1), California (2), and Alaska (3).

High prevalences (up to 83 percent) of *Diphyllobothrium* spp. have been recorded in Eskimo communities throughout northern Canada and Alaska (Arh, 1960; Rausch et al., 1967). In each instance, the high prevalence was linked to the common practice of consuming large amounts of raw fish. The lack of symptoms exhibited among infected persons was also striking (Arh, 1960).

As the popularity of sushi, sashimi, ceviche, and similar dishes increases, so does the risk of diphyllobothriasis. In 1980, the CDC documented an increase in the number of cases of diphyllobothriasis in the United States associated with the consumption of raw or poorly cooked salmon shipped from Alaska to the western United States and Hawaii (Ruttenber et al., 1984). Alaska supplies 75% of the salmon consumed in the United States (CDC, 1981). Between the years 1977 and 1981, 835 cases of diphyllobothriasis were reported in the United States (Ruttenber et al., 1984). As of 1982, the CDC stopped tracking the prevalence of diphyllobothriasis through the use of niclosamide. In 1982 niclosamide was licensed for general distribution whereas prior to that time it was only available through the CDC.

PATHOGENESIS

The pathogenesis of intestinal cestodiasis is poorly understood but the primary pathology is probably due to toxemia resulting form the absorption of tapeworm metabolites.

CLINCAL MANIFESTATIONS AND COMPLICATIONS

Symptoms associated with intestinal cestodiasis are vague, ill-defined, and numerous. Most infections are light and tend to be asymptomatic. When present, symptoms may include diarrhea, abdominal discomfort, anorexia, nausea, weight loss, flatulence, headache, and progressive weakness; the first three listed are most commonly observed.

Massive infections resulting from autoinfection and disseminated infection resulting from metastatic spread of cysticercoids is a concern in immunocompromised patients infected with *H. nana*, but this potential has yet to be conclusively documented.

Although pernicious anemia has long been associated with *D. latum* infection (Nyberg, 1958), this phenomenon appears to be restricted to populations of people in Finland and does not appear to be a feature of diphyllobothriasis in the Americas.

DIAGNOSIS

Diagnosis of intestinal cestodiasis is made by observation of eggs or proglottids in the stool. In the case of *Hymenolepis* spp., diagnosis is made by observation of eggs in the stool. Hamrick (1990) provides an excellent, concise and well-illustrated guide to distinguish between the eggs of *H. nana* and *H. diminuta*. Although differentiation between these two is not of clinical significance since treatment is the same, differentiation does provide important epidemiological information. The presence of eggs in the stool is also diagnostic for diphyllobothriasis.

In the case of *D. caninum*, *Mesocestoides* spp., and often *Diphyllobothrium* spp., diagnosis is made by observation of proglottids in feces. Proglottids of *D. caninum* are ovoid and variable in size, ranging from 8 to 23 mm long (Marty and Neafie, 2000b). The proglottids may be actively moving about and somewhat resemble cucumber seeds (Figure 1). As the proglottids become desiccated and egg packets are released, the proglottids shrivel and dry out until they resemble uncooked rice (Hendrix, 1995). Proglottids may be positively identified microscopically by the presence of dual, lateral genital pores. Proglottids of *Mesocestoides* spp.

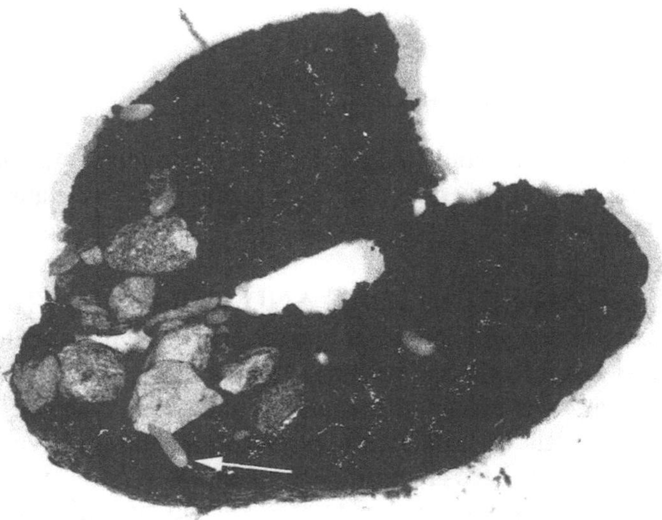

Figure 1. Proglottids of Dipylidium caninum *in cat feces. Courtesy Dr. Thomas Nolan.*

have a mid-ventral genital pore and exhibit a distinct parauterine organ. Gravid proglottids of both *D. caninum* and *Mesocestoides* spp. tend to be longer than they are wide, whereas senile proglottids of *Diphyllobothrium* sp. are wider than long and exhibit a distinct rosette shaped uterus located near the mid-ventral genital pore.

TREATMENT AND CONTROL MEASURES

Hymenolepiasis

The drug of choice for the treatment of hymenolepiasis is praziquantel, given as a single dose of 25 mg/kg for both children and adults.

All cereal and grain products should be tightly sealed and any such products found to be infested with beetles or beetle larvae should be immediately discarded. Given the high prevalence of *H. nana* infection in pet store rodents, newly acquired pet rodents should be screened by a veterinarian for infection with *H. nana.* They should be treated with praziquantel, if infected, before children are allowed to come in contact with them. Hillyer and Quesenberry (1996) recommended 11.4 mg/kg given orally or injected intramuscularly or subcutaneously in 2 doses 10 days apart. Because praziquantel is not made specifically for dosing to rodents, it may be necessary to grind Droncit® (Bayer) tablets and suspend the desired amount of praziquantel in glycerine to prepare a suspension for oral administration. Feces should be examined weekly for eggs and the rodents (and their feces) should be kept away from children and other rodents until stools are negative for *H. nana* eggs.

Dipylidiasis and Diphyllobothriasis

The drug of choice for these is praziquantel (Medical Letter, 2002) given as a single dose at 10 mg/kg for both children and adults. Alternatively, a single dose of 2 g of niclosamide may be given to adults. The pediatric dosage for niclosamide is a single dose of 50 mg/kg.

For prevention of dipylidiasis, dogs and cats should be routinely examined by a veterinarian and treated with praziquantel if found to be infected with *D. caninum*. Infection of humans or pets with *D. caninum* is indicative of flea infestation that should be controlled.

Both freshwater and marine fish should be thoroughly cooked to kill plerocercoids of *Diphyllobothrium* spp. Infection from eating fresh fish can be prevented by cooking until a temperature of 56 °C is achieved throughout for five minutes. Freezing at -18 °C for 24 hours or -10 °C for 72 hours will render plerocercoids inviable. Exposure of fish to a brine solution before cooking should kill plerocercoids as long as the salt concentration is high enough, the fillet is thin enough, and enough time is given in the brine solution (CDC, 1981). Infected pets should also be treated with praziquantel.

Mesocestodiasis

With only six human cases of mesocestodiasis reported, there is obviously little information available concerning treatment. In view of the excellent general results achieved with praziquantel and niclosamide in the treatment of cestodiasis, praziquantel given orally in a single dose of 25 mg/kg is recommended for children or adults. Alternatively, niclosamide in a single dose of 2 g for adults or a single dose of 50 mg/kg for children might be considered. In view of the lack of precedent, vigilant follow-up is warranted.

In the past, quinacrine has been suggested to be the most effective drug in the treatment of mesocestodiasis (Gutierrez, 1978) but has also been shown to be ineffective (Gleason et al., 1973). In view of the excellent cestocidal effects generally obtained with niclosamide and praziquantel, the use of quinacrine for the treatment of tapeworm infections is not recommended. Hutchinson and Martin (1980) claimed that treatment with paromomycin sulfate effected resolution of an infection with *Mesocestoides* sp. in a 17-month-old male. Gleason et al. (1973) reported that a single 1 g dose niclosamide given orally to a 27 month old male resulted in resolution of an infection with *Mesocestoides* sp. Schultz et al. (1992) reported successful treatment of mesocestodiasis in a 22-month-old female with 1 g of niclosamide orally, followed by a second identical dose 2 hours later.

REFERENCES

Arh I. 1960. Fish tapeworm in Eskimos in the Port Harrison area, Canada. *Can J Pub Health* 51: 268-271.

Blagburn BL. 2001. Prevalence of canine and feline parasites in the United States. *Compend Contin Educ Pract Vet* (Suppl.) 23, No. 6(A): 5-10.

CDC. 1981. Diphyllobothriasis associated with salmon-United States. *MMWR* 30: 331-338.

Chandler AC. 1942. First record of a case of human infection with tapeworms of the genus *Mesocestoides*. *Am J Trop Med Hyg* 22: 493-497.

Chappell CL, Enos JP, Penn HM. 1990. *Dipylidium caninum*, an underrecognized infection in infants and children. *Pediatric Inf Dis J* 9: 745-747.

Conti L, Lieb S, Liberti T, et al. 1995. Pet ownership among persons with AIDS in three Florida counties. *Am J Publ Health* 85: 1559-1561.

Crompton DWT. 1999. How much human helminthiasis is there in the world? *J Parasit* 85: 397-403.

Desowitz RS. 1981. *New Guinea Tapeworms and Jewish Grandmothers*. New York: W. W. Norton and Co. 224 pp.

Despommier DD, Gwadz RW, Hotez PJ, Knirsch CA. 2000. *Parasitic Diseases*, 8th ed. New York, NY: Apple Trees Productions, LLC. 345 pp.

Duclos LM, Richardson DJ. 2000. *Hymenolepis nana* in pet store rodents. *Comp Parasit* 67: 197-201.

Flores EC, Plumb SC, McNeese MC. 1983. Intestinal parasitosis in an urban pediatric clinic population. *Am J Dis Child* 137: 754-756.

Gleason NN, Healy GR. 1967. Report of a case of *Mesocestoides* (Cestoda) in a child in Missouri. *J Parasit* 53: 83-84.

Gleason NN, Kornblum R, Walzer P. 1973. *Mesocestoides* (Cestoda) in a child in New Jersey treated with niclosamide (Yomesan®). *Am J Trop Med Hyg* 22: 757-760.

Guarner J, Matilde-Nava T, Villaseñor-Flores R, Sanchez-Mejorada G. 1997. Frequency of intestinal parasites in adult cancer patients in Mexico. *Arch Med Res* 28: 219-222.

Gutierrez Y, Buchino JJ, Schubert WK. 1978. *Mesocestoides* (Cestoda) infection in children in the United States. *J Pediatr* 93: 245-247.

Hamrick HJ, Bowdre JH, Church SM. 1990. Rat tapeworm (*Hymenolepis diminuta*) infection in a child. *Pediatr Infect Dis J* 9: 216-219.

Hendrix CM. 1995. Helminthic infections of the feline small and large intestines: Diagnosis and treatment. *Vet Med* May 1995: 456-472.

Heyneman D. 1962. Studies on helminth immunity: I. Comparison between lumenal and tissue phases of infection in the white mouse by *Hymenolepis nana*. *Am J Trop Med Hyg* 11: 46-63.

Hillyer EV, Quesenberry KE. 1996. *Ferrets, Rabbits, and Rodents: Clinical Medicine and Surgery*. Philadelphia PA: W. B. Saunders Co. 384 pp.

Hutchinson WF, Martin JB. 1980. *Mesocestoides* (Cestoda) in a child in Mississippi treated with paromycin sulfate (Humatin®). *Am J Trop Med Hyg* 29: 478-479.

Isaak DD, Jacobson RH, Reed ND. 1977. The course of *Hymenolepis nana* infections in thymus-deficient mice. *Intl Arch Allergy Appl Immnol* 55: 504-513.

Kappus KK, Juranek DD, Roberts JM. 1991. Results of testing for intestinal parasites by state diagnostic laboratories, United States, 1987. *MMWR* 40: 25-47.

Lucas SB, Hassounah OA, Doenhoff M, Muller R. 1979. Aberrant form of *Hymenolepis nana*: Possible opportunistic infection in immunosuppressed patients. *Lancet* Dec 22/29: 1372-1373.

Lucas SB, Hassounah O, Muller R, Doenhoff MJ. 1980. Abnormal development of *Hymenolepis nana* larvae in immunosuppressed mice. *J Helminthol* 54: 75-82.

Maki J, Yanagisawa T. 1987. Infectivity of *Hymenolepis nana* eggs from faecal pellets in the rectum of mice. *J. Helminthol* 61: 341-345.

Marty AM, Neafie RC. 2000a. Hymenolepiasis and miscellaneous cyclophyllidiases. In: Meyers WM, Neafie RC, Marty AM, Wear DJ (Eds.), *Pathology of Infectious Diseases*, Volume I, *Helminthiases*. Washington, DC: Armed Forces Institute of Pathology. pp.197-214.

Marty AM, Neafie RC. 2000b. Dipylidiasis. In: Meyers WM, Neafie RC, Marty AM, Wear DJ (Eds.), *Pathology of Infectious Diseases*, Volume I, *Helminthiases*. Washington, DC: Armed Forces Institute of Pathology. pp. 137-144.

Medical Letter. 2002. Abramowicz M (Ed.). New Rochelle, NY: The Medical Letter, Inc. 12 pp.

Nyberg W. 1958. Absorption and excretion of vitamin B_{12} in subjects infected with *Diphyllobothrium latum* and in non-infected subjects following oral administration of radioactive B_{12}. *Acta Haemat* 19: 90-98.

Okamoto K. 1969. Effect of cortisone on acquired resistance to *Hymenolepis nana* in mice. *Jap J Parasitol* 18: 591-594.

Rausch RL, Scott EM, Rausch VR. 1967. Helminths in Eskimos in western Alaska, with particular reference to *Diphyllobothrium* infection and anaemia. *Trans Royal Soc Trop Med Hyg* 61: 351-357.

Rembiesa C, Richardson DJ. 2003. Helminth parasites of the house cat, *Felis catus*, in Connecticut, U.S.A. *Comp Parasit* 70: In press.

Roberts LS, Janovy J Jr. 2000. *Gerald D. Schmidt and Larry Robert's Foundations of Parasitology*, 6th ed. Boston, MA: McGraw Hill Co., Inc., 670 p.

Ruttenber AJ, Weniger BG, Sorvillo F, et al. 1984. Diphyllobothriasis associated with salmon consumption in Pacific coast states. *Am J Trop Med Hyg* 33: 455-459.

Schultz LJ, Roberto RR, Rutherforde GW III, et al., 1992. *Mesocestoides* (Cestoda) infection in a California child. *Pediatric Inf Dis J* 11: 332-334.

Turner JA. 1962. Human dipylidiasis (dog tapeworm infection) in the United States. *J Pediatrics* 61: 763-768.

Turner JA. 1975. Other cestode infections. In: Hubbert JA, McCulloch WF, Schnurrenberger PR (Eds.), *Diseases Transmitted from Animals to Man*, 6th ed. Springfield, IL: Charles C. Thomas. pp. 708-744.

OTHER NOTEWORTHY ZOONOTIC HELMINTHS

Dennis J. Richardson

Quinnipiac University, Hamden, Connecticut

The *Merck Veterinary Manual* (1998) lists over 150 zoonotic diseases. Helminth zoonoses constitute nearly a third of that list. There are as many helminth zoonoses listed as rickettsial, bacterial, and protozoal diseases combined. The following is an overview of some lesser-known zoonotic helminthiases. Because they are not common, many of these diseases are underdiagnosed or often misdiagnosed.

Determining which zoonotic helminthiases to exclude has been a difficult task. Two important diseases that were excluded are strongylodiasis and fascioliasis. Strongyloidiasis is transmitted primarily by non-zoonotic means and although fascioliasis is a zoonotic disease, it very rarely occurs in humans in North America. These will be briefly discussed here. The etiologic agent of strongylodiasis is the nematode *Strongyloides stercoralis*. Various estimates place the number of human infections at 70-200 million people worldwide (Genta, 1989; Crompton, 1999; Marquardt et al., 2000). An estimated 400,000 humans are infected in the United States, with prevalences as high as 3-4 percent among school children in the southeastern United States (Berk et al., 1987). Dogs and cats also may be infected with *S. stercoralis.* Although zoonotic transmission does occur (Georgi and Sprinkle, 1974), it plays a minimal epidemiological role in this disease (Pawlowski, 1989). Grove (1989) gives an excellent account of strongylodiasis.

Fascioliasis, caused by the sheep liver fluke, *Fasciola hepatica* occurs in herbivorous mammals, particularly sheep and cattle, that serve as reservoirs for the estimated 2.4 million human cases worldwide, although this may be a gross underestimate (Mas-Coma et al., 1999). Although *F. hepatica* is endemic in cattle in the United States, with prevalences ranging from 6 to 53 percent (Dalton, 1999), only one autochthonously acquired human case has been reported in the past 40 years (Norton and Monroe, 1961). Fifty-three human cases have been reported from Mexico (Mas-Coma et al., 1999). Dalton (1999) provides an excellent account of fascioliasis.

Other zoonotic diseases were excluded because they only occur within limited geographic areas even though they may be regionally important. Salmon-poisoning is one interesting example. This disease, caused by the rickettsial organism, *Neorickettsia helminthoeca*, is transmitted by the intestinal fluke, *Nanophytes salmincola* and may cause enteritis in humans and dogs in the Pacific northwest. Raccoons serve as the primary reservoir for this disease. Milleman and Knapp (1970) and Eastburn et al., (1987) provided good accounts of this disease. Metorchiasis, caused by infection with the liver fluke, *Metorchis conjunctus*, is another such disease. This liver fluke causes a hepatitis similar to that caused by *Opisthorchis sinensus*. Piscivorous (fish eating) carnivores serve as reservoirs throughout northern North America and southward along the Atlantic seaboard to as far south as South Carolina. Humans and dogs may contract the infection by the ingestion of metacercaria in fish. Isolated human cases have been reported throughout Canada. The white sucker, *Catostomus commersoni* appears to be particularly important in the epidemiology of metorchiasis. MacLean et al. (1996) reported an outbreak of acute metorchiasis among 19 people in Montreal who acquired the infection through sashimi prepared from *C. commersoni*. The list of such zoonoses of "minor" importance is extensive. As our knowledge of zoonotic diseases increases and we find that the true importance of these "minor" zoonotic diseases has been underestimated, the significance of the phrase "this wormy world," coined by Stoll (1947) becomes increasingly prophetic.

SPARGANOSIS

Etiologic Agent: *Spirometra mansonoides* in the cestode order Pseudophyllidea

Clinical Manifestations: Skin nodule due to the formation of a subcutaneous granulomatous nodule surrounding the worm

Complications: May occasionally occur in the eye or brain with serious consequences.

Mode of Transmission: Ingestion of infected copepods from freshwater ponds or streams.

Diagnosis: Detection of worm when removed from subcutaneous capsule, although immunodiagnostic tests may be utilized.

Drugs of Choice: None; treatment consists of surgical removal.

Reservoir Hosts: Cats and dogs

Control Measures: Avoid drinking untreated or unfiltered water from natural sources such as ponds and streams. Avoid ingestion of, or preparation of poultices with, raw or undercooked second intermediate or paratenic hosts, especially snakes.

Etiology/Natural History

In humans, the condition of being infected with a plerocercoid of *S. mansonoides* is termed sparganosis. *Spirometra mansonoides* exhibits a typical pseudophyllidean life cycle. It is similar to that of *Diphyllobothrium* spp. with a copepod first intermediate host. Many mammals (including humans) and reptiles are capable of serving as second intermediate or paratenic hosts. Cats and dogs serve as definitive hosts.

Epidemiology

Definitive hosts of *S. mansonoides* are cats and dogs. Adult *S. mansonoides* have been sporadically reported throughout the Eastern United States. *Spirometra mansonoides* has been anecdotally reported as common in dogs and cats in the gulf coastal states (Hendrix, 1995). *Spirometra mansonoides* utilizes several species of mammals and reptiles as second intermediate and paratenic hosts in the southeastern United States (Daly, 1982). Water snakes in the genus *Nerodia* appear to be among the most important second intermediate hosts with prevalences of up to 90 percent being reported in Florida (Daly, 1982). Keeling et al. (1993) found 40 percent of raccoons in southern Florida to be infected with plerocercoids of *S. mansonoides*. It was suggested that the raccoon may play an important role in the maintenance of a sylvatic cycle of *S. mansonoides* by serving as a source of infection for Florida panthers and bobcats, both of which are common definitive hosts of *S. mansonoides* in Florida.

Well over 50 cases of human sparganosis have been reported (Daly, 1982). The majority has been restricted to the southeastern United States, although cases have been reported as far up the Atlantic coast as New York. Louisiana, Arkansas, Florida, and Texas account for about 60 percent of cases reported in the United States. Although there is no clear picture of the prevalence of *S. mansonoides* in domestic dogs and cats, the high prevalence of this parasite in other animals in the southeast (especially Florida) and the preponderance of human cases reported from this region supports the assertion of Hendrix (1995) that this is a common parasite of pets in Florida.

The most common cause of human *S. mansonoides* infection is the accidental ingestion of infected copepods while drinking from freshwater ponds and streams. Infections also may be acquired by ingesting raw or poorly cooked second intermediate or paratenic hosts, including frogs and snakes, or by the use of these animals in the preparation of poultices for application to ulcers, wounds, or sore eyes (Marquardt et al., 2000). The latter means of transmission may play an important role in the epidemiology of this disease in other parts of the world but are probably epidemiologically insignificant in the United States with the possible exception of certain immigrant populations.

Pathogenesis

The worm migrates through the subcutaneous tissue. The initial migratory period is painless and virtually no pathology or symptoms occur (Marty and Neafie, 2000a). The larvae usually subsequently lodge in subcutaneous tissue or muscle where they die and elicit the production of a subcutaneous nodule very similar to that associated with cutaneous filariasis.

Clinical Manifestations and Complications

Sparganosis is characterized by "a painful, inflamed, subcutaneous nodule which may have a history of migration due to movements of the worm" (Daly 1982). Although nodules usually occur in the subcutaneous tissue, they may occur anywhere in the body including the breast, orbit, urinary tract, pericardium, and central nervous system (Marty and Neafie, 2000a). Worms may be long lived, with survival of up to 30 years reported in humans.

Diagnosis

Sparganosis is diagnosed by gross observation of the worm after removal from a subcutaneous nodule. Immunodiagnostic tests are now available for the diagnosis of sparganosis and diagnostic imaging may also be useful. Marty and Neafie (2000a) gave an excellent account of the pathology and diagnosis of sparganosis.

Treatment and Control Measures

The standard treatment of sparganosis is surgical removal of the parasite. If the anterior end of the parasite is not removed there may be regeneration. It therefore is essential that the parasite be completely removed. Infection may be prevented by not drinking, or preparing food with, untreated or unfiltered water from natural sources such as ponds and streams, and not eating or using poultices of raw or undercooked potential second intermediate or paratenic hosts, especially snakes.

CERCARIAL DERMATITIS

Etiologic Agents: Cercariae of the trematode family Schistosomatidae
Clinical Manifestations: Pruritic papular/papulovesicular dermatitis
Complications: Secondary bacterial infection
Mode of Transmission: Penetration of skin by fercocercus cercariae
Diagnosis: Observation of characteristic dermatitis with history of likely exposure to schistosomatid cercariae
Drugs of Choice: Antihistamines and/or antipruritics
Reservoir Hosts: Primarily waterfowl but other birds and mammals as well
Control Measures: Post warning signs in areas of high incidence of cercarial dermatitis; avoid waters likely to contain schistosomatid cercariae

Etiology/Natural History

Cercarial dermatitis is a cosmopolitan disease caused by penetration of the skin by furcocercus (forked-tail) cercariae that normally infect waterfowl and less commonly, other birds and mammals. This condition is associated with both freshwater and marine environments. The presence of the worms elicits an inflammatory reaction leading to cercarial dermatitis although the parasites cannot mature in humans and are quickly destroyed within the dermis. As adults, the etiologic agents of cercarial dermatitis reside in the mesenteric blood vessels of their definitive host where they mate. Eggs are passed in the feces of the definitive host, then hatch and release a ciliated, free-swimming miracidium that penetrates a snail intermediate host. The parasites ultimately develop into cercariae that escape from the snail. The free-swimming cercariae infect a potential definitive host by burrowing through the skin. Within the appropriate definitive host, the worm enters the circulatory system and ultimately migrates to the mesenteric blood vessels, completing the life cycle. Several species of schistosomes utilize a multitude of waterfowl definitive hosts and molluscan intermediate hosts.

Species within the schistosomatid genera *Trichobilharzia*, *Gigantobilharzia*, *Ornithobilharzia*, *Microbilharzia*, and *Heterobiliharzia* have been associated with cercarial dermatitis (Roberts and Janovy, 2000). Snails of the genera *Lymnea*, *Stagnicola*, *Physa*, *Nassarius*, *Planorbis* and *Bulinus* (Marquardt et al., 2000) have been implicated as intermediate hosts for etiologic agents of cercarial dermatitis.

Epidemiology

Cercarial dermatitis can result from exposure to cercariae in both freshwater and marine environments. The common name given to cercarial dermatitis is swimmer's itch. The names claim digger's itch and sea-bather's itch are sometimes applied to marine forms of the disease. In the United States, cercarial dermatitis has been reported from 27 states (Loken et al., 1995) with the greatest number of occurrences being reported from the central and upper Midwest. Marine forms of the disease have been reported form both the United States' Atlantic and Pacific coasts. Cercarial dermatitis is regionally common (Loken et al., 1995) with outbreaks occasionally being reported (CDC, 1982, 1992). The risk for contraction of cercarial dermatitis should be considered in any aquatic environment with relatively high densities of snails and waterfowl, even if only seasonally. Cercarial dermatitis is most prevalent during warmer months but cercaria shedding is been observed as late as November in southern Michigan (Hoeffler, 1974). Areas of shallow water and enhanced vegetation provide good snail habitat and thus enhanced likelihood of the presence of cercariae.

Pathogenesis

Pathology consists of a cutaneous inflammatory response to the presence of schistosomatid cercariae.

Clinical Manifestations and Complications

Just after exposure, a tingling sensation lasting up to an hour may be noted as cercariae penetrate the skin. This phase may be accompanied by a macular eruption and occasionally erythema and/or urticaria (Hoeffler, 1974). Primary itching subsides and the lesions associated with this initial phase may resolve. Later, usually 10-15 hours, an intensely pruritic papular eruption appears. Papules may become vesicular, generally after the second or third day. The rash may persist for days to months but always resolves spontaneously (Hoeffler, 1974; CDC, 1992). Prior sensitization may exacerbate sequelae because of hypersensitivity.

Diagnosis is by observation of characteristic dermatitis with a history of likely exposure to cercariae. The differential diagnoses include contact dermatitis, poison ivy, poison oak, poison sumac, and insect bites (Hoeffler, 1974).

Secondary bacterial infection may occur. Scratching causes tissue damage increasing the likelihood of secondary infection.

Treatment and Control Measures

Symptomatic treatment of dermatitis includes antihistamines and topical antipruritic compounds to relieve inflammation and itching (CDC, 1992).

CUTANEOUS LARVA MIGRANS (CREEPING ERUPTION)

Etiologic Agents: Various hookworm larvae, most commonly *Ancylostoma braziliense* and *Ancylostoma caninum*

Clinical Manifestations: Dermatitis characterized by a progressive, narrow, raised, serpiginous, pruritic lesion

Complications: Rarely invasion of lungs and/or muscles by larvae.

Mode of Transmission: Humans become infected by penetration of infective L3 larvae in fecally contaminated soil.

Diagnosis: Presumptive, based on the presentation of a characteristic serpiginous lesion, especially with a history of exposure to potentially fecally-contaminated soil

Drug of Choice: Thiabendazole ointment

Reservoir Hosts: Many mammals; primarily dogs and cats

Control Measures: Wear shoes, avoid contact of skin with soil that may have been contaminated with animal feces, such as at beaches and public parks. Animal control; leash and litter laws should be strictly enforced in public areas. Sandboxes should be covered when not in use.

Etiology/Natural History

The larvae of a variety of hookworms that normally occur as adults in non-human mammals may penetrate the skin of humans, leading to a progressive, extremely pruritic dermatitis known as creeping eruption. In their normal definitive host, usually dogs or cats for *A. braziliense* and dogs for *A. caninum*, adult worms live and mate in the small intestine. Eggs are passed in the feces and hatch in warm, moist soil. A larva hatches from the egg and feeds on bacteria; after two additional molts, the third stage (L3) larva is infective to a definitive host. Entrance is gained by penetrating the skin. In the normal definitive host, the larval worms enter the circulatory system and undergo a pulmonary migration that ultimately leads to the small intestine where they mature into adults. These L3 larvae also may penetrate the skin of humans and wander within the subcutaneous tissue for a period of weeks to months causing the condition known as cutaneous larva migrans, or creeping eruption. A few other helminths in the genera *Strongyloides* and *Gnathostoma* cause similar conditions in humans. *Strongyloides procyonis*, a raccoon helminth, causes a condition in humans known as "duck hunters' itch" (Jelinek et al., 1994).

Epidemiology

Creeping eruption is most common in the tropics and subtropics because infective, L3 larvae thrive in moist, warm soil; however, creeping eruption also occurs throughout North America. In the United States, creeping eruption is most common in the Southeast. Public areas such as beaches and public parks where pets and people congregate are common sites where humans become infected. When dog or cat feces containing hookworm eggs have contaminated the soil, human infection may occur if skin comes into contact with soil that contains L3 larvae. Such activities as walking barefoot, playing in contaminated sandboxes, and sunbathing provide opportunity for infection. Workers with occupations that bring them in frequent close contact with soil, such as gardeners, plumbers, and electricians, are prone to infection (Muhleisen, 1953).

Pathogenesis

Creeping eruption presents as a dermatitis. It begins with a pruritic erythmatous papule that appears at the site of penetration within hours of exposure. The parasite then begins emigrating between the stratum germinativum and the stratum corneum and produces a raised, firm serpiginous lesion (CDC, 1981). Surrounding tissue is edematous and erythmatous. The dermatitis is characterized by intense pruritis. The larva is typically one to two cm ahead of the lesion (Jelinek et al., 1994). The condition may persist from weeks to months, until the larvae die, degenerate, and are absorbed. Inflammation in response to larval secretions is the primary pathology (Despommier et al., 2000). Feet, buttocks, ano-genital region, and legs are the most commonly affected sites. (Jelinek et al., 1994).

Clinical Manifestations and Complications

The primary symptom of creeping eruption is intense pruritis. Scratching may induce secondary bacterial infection. In rare instances, larvae may invade the lungs, causing pulmonary eosinophilic infiltrate with fever, cough, and chest pain (Little et al., 1983; Jelinek et al., 1994). Muscle tissue also may be invaded. In rare instances this leads to focal myositis characterized by eosinophilic inflammation and degeneration of fibers with muscle pain and swelling (Little et al., 1983). Also, in rare instances, the dog hookworm, *A. caninum*, may establish enteric infection characterized by eosinophilic inflammation with acute abdominal pain (Croese et al., 1994).

Diagnosis

Diagnosis is usually presumptive based on observation of the characteristic serpiginous lesion (Figure 1). Epidermal sequela mimics that of acute allergic contact dermatitis from poison ivy, poison oak, or poison sumac (Sulica et al., 1988). Because the larva travels "in front" of the observable lesion, locating the larvae is very difficult and renders biopsy or surgical intervention unsuccessful (Jelines et al., 1994).

Figure 1. Typical serpiginous lesion associated with creeping eruption on the side of a foot. Courtesy Ron Neafie, AFIP

Treatment and Control Measures

The drug of choice is thiabendazole ointment applied topically to the lesions. In severe cases, albendazole (400 mg daily x 3d) or ivermectin (200 μg/kg daily x 1-2 d) may be given orally to children and adults in addition to thiabendazole ointment (*The Medical Letter*, 2002). Prednisone has been given topically and systemically to reduce inflammation in severe cases

(CDC, 1981). Ethyl chloride spray provides quick but short-lived relief of pruritis (Despommier et al., 2000).

Exposed skin should not come in contact with fecally contaminated soil; this is especially true in public areas such as beaches and parks where pets are allowed to defecate. Shoes should always be worn. When sunbathing, skin should not be exposed to potentially contaminated sand or soil. This is especially important when vacationing in tropical or subtropical locales (CDC, 1981). Animal control should be exercised. If pets are permitted in public areas, leash and litter laws should be stringently enforced. Sandboxes should be covered when not in use.

DIROFILARIASIS

Etiologic Agents: Filarial worms including *Dirofilaria tenuis, Dirofilaria ursi, Dirofilaria subdermata,* and *Dirofilaria immitis*

Clinical Manifestations: Subcutaneous nodules, tender to firm, with or without pain and/or pruritis. Lesions may be migratory. Pulmonary dirofilariasis presents as a spherical subplerual pulmonary infarct with a self-limiting granulomatous inflammatory response. Radiographically, these appear as "coin lesions."

Complications: Typically none. Pulmonary lesions associated with *D. immitis* are usually mistaken for primary or metastatic lung tumors. Occasionally, nodules associated with *D. tenuis*, *D. ursi*, and *D. subdermata* will present as breast lumps.

Mode of Transmission: Humans become infected by inoculation of microfilariae from the salivary glands of a mosquito (for *D. tenuis* and *D. immitis*) or black fly (for *D. ursi*).

Diagnosis: Detection of the worm in histological section following surgical excision of the nodule.

Drug of Choice: None

Reservoir Hosts: Raccoons, bears, porcupines, and dogs for *D. tenuis, D. ursi, D. subdermata,* and *D. immitis* respectively

Control Measures: Precautions to avoid mosquito and black fly "bites;" implementation of aggressive control programs for *D. immitis* infections in dogs.

Etiology/Natural History

As adults, *D. tenuis, D. ursi,* and *D. subdermata* live in the subcutaneous tissue of raccoons, bears, and badgers, respectively. Microfilariae produced by adult female worms enter the circulatory system and are ingested by a feeding mosquito or sandfly. Mosquitos serve as the vector for all species of *Dirofilaria* except for *D. ursi*, which uses black flies, especially *Simulium venustum*, as a vector (Marty and Neafie, 2000b). *Aedes taeniorhynchus* appears to be an especially common vector of *D. tenuis* (Pung et al., 1996).

The life cycle of *D. subdermata* is not known. *Dirofilaria tenuis* is common in parts of the southeastern US and occurs as far north as Maryland. *Dirofilaria tenuis* is particularly abundant in Florida (Isaza and Courtney, 1988). *Dirofilaria ursi* and *D. subdermata* occur parapatrically and cannot be differentiated based on histological sections. This observation has lead Beaver et al. (1987), to suggest that human infections with either *D. ursi* or *D. subdermata*, be categorized as infection with *D. ursi*-like worms. Human infections with *D. ursi*-like worms have been reported from Washington, across Canada and upper Michigan to Quebec, northern Vermont, and upstate New York. Earlier reports of probable *D. tenuis* as a cause of infections in Ontario (Anderson et al., 1968), Winnipeg (Conly et al., 1984), Michigan (Payan, 1978), New York (Davies et al., 1973) and Vermont (Christie, 1972) were attributed by Beaver et al. (1987) to *D. ursi*-like worms.

Canine heartworms, *D. immitis*, live as adults in the right ventricle of the heart and pulmonary artery. Adult female worms pass microfilariae into the circulatory system that may be ingested by a feeding mosquito. Microfilariae develop within the mosquito into an infective larval stage. While taking a blood meal, mosquitoes may transmit infective larvae from their mouthparts to a dog or other host. The larvae undergo further development in the subcutaneous, adipose, and muscular tissue of the dog. After 90 days, worms arrive at the right ventricle or pulmonary artery where they mature into adults and mate (Roberts and Janovy, 2000; Marquardt et al., 2000). *Dirofilaria immitis* is broadly distributed throughout North America and many species of mosquito may serve as intermediate hosts.

Epidemiology

Humans become infected with *Dirofilaria* spp. when larvae are transmitted by an infected feeding mosquito (black fly for *D. ursi*); however, *Dirofilaria* spp. do not mature in humans. Human infections with *D. ursi*-like worms have been sporadically reported throughout Canada, and the northern and northeastern United States. Infections with *D. tenuis* are restricted to the southeastern United States, with 80 percent of cases reported from South Florida, especially Collier and Brevard counties. Infections with *D. tenuis* may be contracted during travel to this region (Collins et al., 1993). Prevalence of human infection depends to a great extent on the prevalence of the parasite in reservoir hosts in conjunction with the density of suitable vectors. For instance, the prevalence of *D. tenuis* in raccoons ranges from 6 percent in north Florida to 21 percent and 45 percent in Brevard and Collier counties, respectively, in south Florida (Isaza and Courtney, 1988). The black salt marsh mosquito, *A. taeniorhynchus*, which has been implicated as a natural intermediate host for *D. tenuis* (Pung et al., 1996), is abundant in southern Florida (Collins et al.,

1993). Outside of southern Florida, the prevalence of *D. tenuis* tends to be relatively low; however, Pung et al. (1996) found that 20 percent of raccoons in portions of coastal and near-coastal southeastern Georgia were infected with *D. tenuis*. Furthermore, 3.4 percent of *A. taeniorhynchus* collected from St. Catherine's Island in southeastern Georgia were infected with *Dirofilaria*-like larvae that were likely *D. tenuis*. Human dirofilariasis has been reported in Georgia. Pung et al. (1996) predicted that human exposure to raccoon filariae might increase as the Georgia coast continues to be developed for commercial and residential use.

Human pulmonary dirofilariasis resulting from infection with the dog heartworm, *D. immitis*, has been reported sporadically throughout the eastern and northeastern United States, primarily east of the Mississippi River. West of the Mississippi River, human infections have been reported in Louisiana, Texas, and California. Infections are most common in the eastern, southeastern, and southern coastal states with Texas, Louisiana, and Florida accounting for over half of the known human infections in the United States (Asimacopoulos et al., 1992). Traditionally, infection in dogs was confined to the eastern and southeastern United States. In recent years there has been an increase in the prevalence of this disease along with a substantial expansion of its geographic range (Asimacopoulos et al., 1992). As of 1992, canine dirofilariasis had been reported from every state except Nevada. Nevertheless, canine infections remain more common in the eastern and southeastern United States where prevalences as high as 45 percent occur (Roberts and Janovy, 2000). Asimacopoulos et al. (1992) pointed out that although human heartworm infection is rare, it is exhibiting an increase in prevalence that parallels the rise in prevalence and extension of the geographic range of canine infection. Alvarez et al. (1990) recorded a *D. immitis* seroprevalence of 5 percent among the general human population in Salamanca, Spain. The prevalence of heartworm infection in dogs in the same area was 35.6 percent (Perez Sanchez et al., 1989), not unlike the prevalence in some areas of the United States. In view of these findings, it seems likely that human infection with *D. immitis* is underdiagnosed. As greater awareness and proper diagnosis of human dirofilariasis emerges, the number of reported cases will probably continue to increase.

Pathogenesis

When infecting a human, *D. tenuis* and *D. ursi*-like worms become encapsulated in a subcutaneous nodule. *Dirofilaria immitis* probably initially locates in the right ventricle of the heart but it cannot mature into an adult and dies. The dead worm is then carried to the lung where it becomes encapsulated.

Human infection with *D. tenuis* presents as a nodule or swelling of variable size, usually 1-3 cm. Lesions may or may not be pruritic or painful and may exhibit varying degrees of erythrema and tenderness. Occasionally, the lesion may be migratory. Such subcutaneous nodules may occur anywhere on the body but are usually found in order of frequency, on the arm or hand, head or neck, leg, and at or near the breast (Beaver and Orihel, 1965). Nodules typically contain a dead nematode and microscopically demonstrate a "florid granulomatous and eosinophilic inflammatory response associated with marked fibrosis of the connective tissue (Collins et al., 1993)."

Human infection with *Dirofilaria immitis* presents as a "spherical, subpleural pulmonary infarct with granulomatous reaction that is almost always a self-limited process, posing no significant threat to human health (Asimacopoulos et al., 1992)." The pulmonary lesion that is recognized radiographically as a "coin lesion" (Collins et al., 1993) is virtually identical in presentation to a primary or metastatic tumor (Asimacopoulos et al., 1992).

Clinical Manifestations and Complications

Human dirofilariasis is usually asymptomatic. The similarity of the lesions to cancerous neoplasia necessitates surgical excision of the nodules, especially for pulmonary lesions and breast lumps (Conly et al., 1984; Gutierrez, 1984). Over half of the cases of pulmonary dirofilariasis are asymptomatic. When present, symptoms are usually mild and may include cough (23-30%), chest pains (17%), eosinophilia (10-15%), hemoptysis (9-10%), cold sweats (10%) and fever (6%) (Asimacopoulos et al., 1992; Roy et al., 1993). Occasionally, extra-pulmonary infections with *D. immitis* occur. Theis et al. (2001) reported a case of an eosinophilic granulomatous nodule associated with *D. immitis* involving the spermatic cord. The nodule was so densely adherent to the vas deferens, testicular artery, and veins that an orchiectomy was performed. Although rare, acute arthritis has been associated with subcutaneous dirofilariasis (Corman, 1987). Collins et al. (1993) reported *D. tenuis* infection of the oral mucosa. Occasionally, *D. tenuis* (syn. *Dirofilaria conjunctivae*) infections may have ocular involvement, usually limited to the subcutaneous tissue of the eyelid and periorbital region, although subconjunctival involvement can occur (Marty and Neafie, 2000b).

Diagnosis

Diagnosis is made by detection of the worm in histological section of a surgically excised nodule.

Treatment and Control Measures

Treatment consists of complete surgical excision of the infected tissue (Collins et al., 1993). Aggressive control programs for canine heartworm infection are warranted. In regard to all dirofilarids, the best way to avoid infection is to avoid being "bitten" by mosquitoes (black flies in the case of *D. ursi*), especially in Southern Florida where *D. tenuis* is common. Precautions include the use of DEET, avoidance of outdoor activities at times of peak mosquito activity, and wearing appropriate clothing to cover as much skin as possible when outdoors.

ZOONOTIC BRUGIAN FILARIASIS

Etiologic Agents: Filarial worms of the genus *Brugia*
Clinical Manifestations: Enlarged lymph node
Complications: Typically none.
Mode of Transmission: Humans likely become infected by inoculation of microfilariae from the salivary glands of a mosquito.
Diagnosis: Observation of worm in histological section following surgical excision of the infected lymph node.
Drug of Choice: None
Reservoir Hosts: Probably raccoons and rabbits; potentially other mammals
Control Measures: Take precautions to avoid mosquito bites

Etiology/Natural History

At least 31 human cases of zoonotic brugian filariasis caused by infection with filarial worms in the genus *Brugia* have been reported since 1962: 8 cases from New York, 3 each from Connecticut, Massachusetts, Rhode Island, and Pennsylvania, 2 from Canada and one each from New Jersey, Michigan, Ohio, Florida, North Carolina, Louisiana, Mississippi, Oklahoma, and California (Orihel and Beaver, 1989; Eberhard et al., 1993; Elenitoba-Johnson et al., 1996; Baird et al., 2000). Two species of brugian filarial worms have been described in North America: *Brugia beaveri* from raccoons in Louisiana (Ash and Little, 1964) and *Brugia lepori* from Louisiana rabbits (Eberhard, 1984). An undescribed species of *Brugia* was reported from a cat in California (Beaver and Wong, 1988). It is not clear if the species responsible for human cases is *B. beaveri*, *B. lepori*, or an undescribed species in the genus *Brugia* (Baird et al., 1986).

Epidemiology

Most infected humans have resided in suburban or semi-rural areas (Baird et al, 1986). This probably is due to close proximity with reservoir hosts and thus infected mosquitoes. It is not clear, which, if either, of the known North American species of *Brugia* is responsible for human infections. Eberhard (1984) reported a prevalence of 70% for *B. lepori* infection in wild rabbits in Louisiana and Eberhard et al. (1991) found that 60% of wild rabbits on Nantucket Island, Massachusetts were infected with *Brugia* sp.(most likely *B. leporis*). Such high prevalences of this parasite in regions of the country where the preponderance of human cases have been reported implicate *B. leporis* as the most likely etiologic agent for zoonotic human brugian filariasis and rabbits as a potentially important reservoir. Likewise, Ash and Little (1964) found 5 of 7 Louisiana raccoons to be infected with *B. beaveri*, implicating these animals as potential reservoirs and this parasite as an etiologic agent in the southeast. Similar microfilariae were reported from a bobcat in Florida (Ash and Little, 1964).

Pathogenesis

Pathology is restricted to the effected lymph node(s), and is minimal. Infected lymph vessels are usually dilated (Baird et al., 1986). Typically only one node is involved and infection is restricted to a single worm, although infections with multiple worms in multiple sites do occur. Worms are often dead and surrounded by a necrotizing granuloma. Thickened vessel walls may occur and if so are often infiltrated with inflammatory cells (Baird et al., 1986). Worms are occasionally found alive, but with a single exception (Simmons et al., 1984), do not achieve patency.

Clinical Manifestations and Complications

American brugian filariasis presents as local lymphadenopathy, usually of a single node. Infected nodes generally are not tender, although they occasionally may be tender and rarely painful. Typically, the lesions are non-erythmatous but in rare instances "streaking" associated with the lympadenitis may be suggestive of cellulitis (Eberhard et al., 1993). Although the infected nodes may occur virtually anywhere on the body, the most common sites of infection are the groin (34%), neck (21%), and axilla (10%).

Pathology is usually limited, to a single node. In the case of an immunodeficient infant in Oklahoma suffering from zoonotic brugian filariasis, the disease progressed to severe lymphadema (Simmons et al., 1984). Interestingly, this is the only case of American brugian filariasis in which patency was observed, evidenced by microfilaremia. Simmons et al. (1984) suggested that the B-cell immunodeficiency in this patient permitted maturation of the worms and concomitant generalized lymphadema. Thus, zoonotic brugian filariasis warrants heightened concern when occurring in an immunocompromised patient, particularly in the case of a B-cell immunodeficiency. This approach is strengthened by the observation made by Greene et al. (1978) of microfilaremia associated with a case of non-brugian filariasis in a teenager in Alabama who was on prednisone therapy for systemic lupus erythematosus.

A Connecticut case involved pulmonary infection with *Brugia* sp. (Orihel and Beaver, 1989), originally identified as *D. immitis* (Levinson et al., 1979). This infection resulted in an infarct with subsequent lesions characteristic of pulmonary filariasis.

Diagnosis
Diagnosis is typically made by examination of worm(s) in histological sections from the surgically excised lymph node. Preoperative diagnoses have included papilloma, lymphoma, and lipoma (Baird et al., 1986).

Treatment and Control Measures
Removal of the infected lymph node or lymphatic tissue affects a complete cure. The only realistic control measure is to avoid being "bitten" by mosquitoes.

ANISAKIASIS

Etiologic Agents: *Anisakis simplex* and *Pseudoterranova decipiens* (syn. *Phocanema decipiens*), and rarely other species of the nematode family Anisakidae
Clinical Manifestations: Symptoms of infection with *A. simplex* may include mild to severe abdominal pain, nausea, vomiting, diarrhea, and fever. Infections with *P. decipiens* are less serious and may be characterized by minor throat irritation, coughing, expectoration, nausea and vomiting.
Complications: Extra-gastrointestinal infection may occur involving the mesentery, lung, spleen, or liver. Infection with *A. simplex* may induce allergic reactions, including urticaria, angioedema, and, in rare instances, anaphylaxis.
Mode of Transmission: Humans become infected by ingesting third stage larvae in raw or poorly cooked fish or squid. Cod, herring, mackerel, salmon, tuna, and yellowtail are particularly important sources of infection.
Diagnosis: Dectection of the worm after it is coughed up or vomited, or detection by endoscopy or immunodiagnostic tests.
Drug of Choice: No drugs have been found to be effective. Treatment consists of endoscopic or surgical removal of the worm(s).
Reservoir Hosts: Various sea mammals including seals, sea lions, walruses, porpoises, and whales
Control Measures: Avoid ingesting raw or poorly cooked fish or squid.

Etiology/Natural History
Although several species of anisakid nematodes may cause disease in humans, *A. simplex* and *P. decipiens* are responsible for North American cases with *P. decipiens* accounting for the majority. Adult worms live and mate in the alimentary canal of marine mammals and release eggs that are

passed in feces. *Anisakis simplex* utilizes porpoises and whales and *P. decipiens* utilizes seals, sea lions and walruses as definitive hosts. Eggs hatch in water thereby releasing second stage (possibly third stage) larvae that are ingested by small crustaceans (especially krill), in which infective third stage larvae may be found. When small fish ingest infected crustaceans, larvae enter the peritoneum or muscle tissue. Fish and squid serve as paratenic hosts. In a type of biological magnification, the larvae may be passed up the food chain from fish to fish leading to the accumulation of a large number of larvae in single large fish. Definitive hosts become infected by ingesting larvae, often in large numbers, that establish and mature in the alimentary canal. Rockfish and salmon are the most common paratenic hosts in North America (Schantz, 1989; Anderson and Lichtenfels, 2000).

Epidemiology

Humans become infected with anisakid nematodes when they ingest infective larvae while eating poorly cooked or raw fish. Fish most commonly involved include salmon, rockfish (Pacific red snapper), herring, mackerel, cod, pollack, tuna, and yellowtail (Schantz, 1989; Deardorff et al., 1991; Anderson and Lichtenfels, 2000). In addition to high-intensity anasakid infections, fish may exhibit high prevalences of infection. Deardorff et al. (1991) found that 91% of Pacific salmon imported into the Hawaiian fish market were infected with *A. simplex*. *Anisakis simplex* and *P. decipiens* occur in both the Atlantic and Pacific oceans but are more common in the Pacific, probably because of higher population densities of sea mammals. The first case of human anisakiasis in North America was reported in 1955 in a patient who acquired the infection while eating lightly smoked herring (Jackson, 1975). Since then, well over 50 cases have been reported. Anisakiasis occurs wherever raw fish is consumed in such dishes as sushi, sashimi, and ceviche. As the popularity of these dishes increases, the incidence of anisakiasis is likely to increase.

Pathogenesis

Infection with *P. decipiens* is usually not serious because the worms have a tendency not to invade the gastric or intestinal mucosa. Worms are usually coughed up or vomited within 48 hours of ingestion. Larvae of *A. simplex*, on the other hand, often penetrate the mucosa of the stomach or intestine, causing moderate to acute gastritis or enteritis respectively (McKerrow et al., 1988). Although the worms do not usually mature in humans, infections may persist for weeks or even months.

Acute gastroenteritis occurring early in the infection may result from a hypersensitivity reaction, described below. The chronic stage of the infection is characterized by extensive inflammation with larvae stimulating massive tissue eosinophilia and disrupting blood vessels . This leads to ulceration, thickening of the mucosa, and hemorrhage (Anderson and Lichtenfels, 2000). As the larva dies, it is eventually contained within a granuloma.

Another interesting aspect of the pathology of anisakiasis is that in many instances, a type I hypersensitivity reaction known as gastroallergic anisakiasis occurs in association with mucosal invasion by *A. simplex*. The hypersensitivity-associated symptoms appear to be a sequelae of an IgE-mediated response, along with recruitment of eosinophils, that is characteristic of a typical helminth-induced T_H2-mediated immune response (Daschner et al., 2000). The onset of symptoms associated with gastric anisakiasis usually occurs within 12 hours of ingestion of worms, whereas hypersensitivity symptoms such as urticaria, angioedema, and occasionally anaphylaxis (Moreno-Ancillo et al., 1997; Del Pozo et al., 1997) usually occur in about half that time (mean 5.3 hours). These symptoms may take as long as 26 hours to appear (Daschner et al., 2000). Alonso et al. (1999) and Sastre et al. (2000) provided evidence suggesting that antigens of live parasites probably cause gastroallergic anisakiasis, based on a lack of response in patients previously sensitized to *A. simplex*, when challenged with frozen or lyophilized larvae of *A. simplex*. Moneo et al. (2000) isolated a 24-kd protein, *Ani s 1*, that occurs exclusively in the excretory glands of *A. simplex*. Consistent with the assertions of Alonso et al. (1999) and Sastre et al. (2000), Moneo et al. (2000) suggested that only during tissue invasion, would enough *Ani s 1* be produced to be strongly allergenic.

Clinical Manifestations and Complications

Symptoms of infection with *P. decipiens* may include irritation or a "tickling sensation" in the pharyngeal region as the worm(s) migrate(s) in the oropharynx or proximal esophagus. This may be accompanied by coughing, and/or nausea and vomiting. The worms are usually coughed up or vomited within 48 hours of ingestion (McKerrow et al., 1988; Schantz, 1989).

Symptoms of infection with *A. simplex* may include moderate to severe abdominal pain, fever, nausea, vomiting, or diarrhea appearing within hours (usually 1 to 12) of eating seafood. Differential diagnoses include acute appendicitis, gastric ulcer, Crohn's disease, or gastrointestinal cancer.

Infections usually are restricted to the gastric and enteric systems, although extra-alimentary sites of infection are occasionally observed, including the mesentery, lung, spleen, pancreas, and liver (Moreno-Ancillo et al., 1997).

Hypersensitivity manifestations of gastroallergic anisakiasis include urticaria, angioedema, and anaphylaxis. Moreno-Ancillo et al. (1997) studied the allergic reactions to *A. simplex* in 23 patients who suffered allergic reactions after seafood ingestion and showed a negative skin prick test for the food and positive skin prick test of *A. simplex* sensitization. Moreno-Ancillo et al. (1997) observed urticaria and/or angioedema in 78 percent of patients with gastroallergic anisakiasis, and anaphylaxis (defined as a life threatening reaction involving more than two organs) in 22 percent. Only 13 percent exhibited gastric symptoms. Thus, it is important to consider *Anisakis* allergy in the differential diagnosis of food allergy. This is especially important because there is evidence suggesting that individuals sensitized to *A. simplex* may safely consume seafood as long as all larvae are killed.

Diagnostic Tests:

Diagnosis of infection with *P. decipiens* is usually made by identification of the worm(s) from expectorated material or vomitus.

Diagnosis of infection with *A. simplex* infection is usually made viewing the worm endoscopically.

Diagnosis of gastroallergic anisakiasis may be made by a skin prick test or by determination of serum IgE antibodies using a CAP-FEIA (Pharmacia Biosystems, Sweden) assay or immunoblotting techniques, each of which utilize whole parasite antigen. Although these tests exhibit 100, 100, and 96 percent sensitivity respectively, they are highly cross-reactive due to their dependence on whole antigen. Lorenzo et al. (2000) described an antigen-capture ELISA using a deglycosylated parasite allergen captured by the monoclonal antibody UA3 that shows promise of higher specificity so that UA3-ELISA may be a valuable tool for confirmation of a positive skin prick test.

A food consumption history is extremely important in the diagnosis of anisakiasis. Gastric symptoms often are not exhibited in patients with gastroallergic anisakiasis. The onset of an IgE-mediated hypersentivity reaction generally is manifested within 2 hours with food allergies. In the case of gastroallergic anisakiasis, it may be as long as 26 hours after ingestion of the infected seafood.

Treatment and control measures

Treatment is usually not required for *P. decipiens* infection. Treatment of *A. simplex* infection usually entails endoscopic or surgical removal of the worm(s).

Fresh fish should always be thoroughly cooked (heated to 65 °C for 10 minutes). Raw fish should be frozen for at least five days at –20 °C before consumption. Cold smoking and brining methods are not reliable in rendering all larvae inviable (Schantz, 1989), thus it should not be assumed that delicacies such as smoked herring are "absolutely safe." The vast majority of cases of anisakiasis are acquired through fish dishes prepared at home. Only rarely are cases associated with restaurants, even those specializing in dishes containing raw fish (Schantz, 1989). This is probably because most fish are thoroughly frozen prior to use in restaurants.

Recent Advances and Contemporary Challenges

Although it has been known for four decades that anisakid nematodes may elicit an allergic response (Smith and Wootten, 1970) the immune response mechanisms against anisakiasis have only begun to be elucidated. To date all such work has been conducted in Europe, particularly in Spain. Studies are needed to evaluate the role of *Anisakis* and potentially *Pseudoterranova* in seafood allergies in North America. It is hypothesized that *Pseudoterranova* will exhibit little if any allergic sequela because it tends not to invade tissues and it appears that tissue invasion is a prerequisite to hypersensitivity reaction for *A. simplex*.

Anisakis allergens have been putatively linked to rheumatic symptoms (Cuende et al., 1998), conjunctivitis (Añibarro and Seoane, 1998) contact dermatitis (Carretero Anibarro et al., 1997), and asthma (Armentia et al., 1998). This system provides a model offering opportunity for further elucidation of the immunobiology of parasitism. It also may help in further investigating the potential role (or lack thereof) of helminth infection as an exacerbating or precluding factor for human asthma (Tullis, 1970; Turner, 1978; Jarrett et al., 1980; Feldman and Parker, 1992; Sharghi et al., 2001). It is particularly intriguing that substantial IgE cross-reactivity was detected between *A. simplex*, three species of dust mites (Johansson et al., 2001) and the German cockroach (Pascual et al., 1997).

ACANTHOCEPHALIASIS

Etiologic Agents: Acanthocephalans (thorny-headed worms), particularly *Moniliformis moniliformis*, *Macracanthorhynchus ingens* and *Macracanthorhynchus hirudinaceus*

Clinical Manifestations: Symptoms may include abdominal pain, diarrhea, tinnitus/dizziness, anorexia, mild fever, abdominal distension, nausea, and vomiting

Complications: Although very rare, intestinal perforation has been reported.

Mode of Transmission: Humans become infected by ingesting cystacanths in infected intermediate hosts which include cockroaches for *M. moniliformis* and beetles for *Macracanthorhynchus* spp.

Diagnosis: Detection of eggs or adult worms in stool

Drug of Choice: Thiabendazole

Reservoir Hosts: Rats for *M. moniliformis*, raccoons for *M. ingens*, and swine for *M. hirudinaceus*

Control Measures: Avoid ingesting raw beetles and cockroaches.

Etiology/Natural History

Of the slightly more than 1100 valid species in the phylum Acanthocephala, the thorny-headed worms, none utilize humans as primary definitive host. Human infections do occur, however, especially with *M. moniliformis* and *Macracanthorhynchus* spp. Adult acanthocephalans live and mate in the small intestine of the definitive host. Intermediate hosts become infected by ingesting parasite eggs passed in feces of the definitive host. Within the intermediate host, the worm develops into a cystacanth. Definitive hosts (including humans) become infected by ingesting cystacanths in an infected intermediate host. *Moniliformis moniliformis* utilizes a rat definitive host and cockroach intermediate host. *Macracanthorhynchus ingens* and *M. hirudinaceus* utilize a raccoon and swine definitive host, respectively. Both *M. ingens* and *M. hirudinaceus* utilize various beetles as intermediate hosts. Natural infections of *M. ingens* have also been reported in wood roaches and a millipede (Schmidt, 1985).

Epidemiology

Most human cases of acanthocephaliasis occur in infants and toddlers, although infection of adults does occur. Acanthocephaliasis has been sporadically reported around the world over the past four decades. Four cases of acanthocephaliasis have been reported in North America, two involving *M. moniliformis* in Florida (Beck, 1959; Counselman et al., 1989) and two involving *M. ingens* in Texas (Dingley and Beaver, 1985; Vienn et al., 2000). Risk of infection increases where there are high population densities of infected rats, raccoons or swine.

Pathogenesis

Attachment of the proboscis results in localized trauma of the intestinal mucosa and submucosa leading to granulomatous infiltration and collagenous encapsulation around the proboscis (Richardson and Barnawell, 1995; Roberts and Janovy, 2000). Nelson and Nickol (1986) found such chronic, fibrotic lesions to be associated with *M. hirudinaceus* in swine and *M. ingens* in raccoons, but less severe in the latter. Regions of focal hemorrhage were associated with *M. hirudinaceus* in swine. Similar gross pathology and histopathology resulting from chronic inflammation apparently occur in infected humans (Pradatsundarasar and Pechranond, 1965; Kliks et al., 1974). Taraschewski et al. (1989) found that lesions of *M. moniliformis* in rats were more superficial than those evoked by *Macracanthorhynchus* spp., involving only the mucosa and tunica propria. This indicates that these worms frequently change attachment sites. Much of the systemic sequela observed probably results from toxemia induced by worm metabolites. Neafie and Marty (2000) gave an excellent account of the pathology of acanthocephaliasis.

Clinical Manifestations and Complications

Of 27 case reports of moniliformiasis presented in the literature (Grassi and Calandruccio, 1888; Beck, 1959; Sahba et al., 1970; Moayedi et al., 1971; Goldsmid et al., 1974; Al-Rawas et al., 1977; Counselman et al., 1989; Prociv et al., 1990; Anosike et al., 2000), 33% were asymptomatic. When present, the most common symptoms were mild to severe abdominal pain (39%), diarrhea (22%), tinnitus/dizziness (22%), and anorexia (17%). Other clinical manifestations noted in 11% of patients were mild fever, abdominal distension, physical retardation, and nausea and vomiting. Too few case reports of infection with *Macracanthorhynchus* spp. exist to provide a summary of symptoms; however, existing information (Pradatsundarasar and Pechranond, 1965; Kliks, et al., 1974; Dingley and Beaver, 1985; Radomyos et al., 1989; Vienn et al., 2000) suggests symptomology similar to that evoked by *M. moniliformis*.

Although rare, intestinal perforation with subsequent peritonitis has been documented (Kliks, 1974; Radomyos et al., 1989). It has been suggested that secondary bacterial infection of lesions resulting from worm attachment may lead to disseminated foci of necrosis (Schmidt, 1972; Kliks et al., 1974).

Diagnosis

The acute abdominal pain associated with acanthocephaliasis may mimic that of appendicitis. Diagnosis is made by observation of worms or eggs (Figure 1) in stool. Adult *M. moniliformis* (Figure 2) exhibit conspicuous pseudosegmentation, sometimes superficially resembling tapeworms.

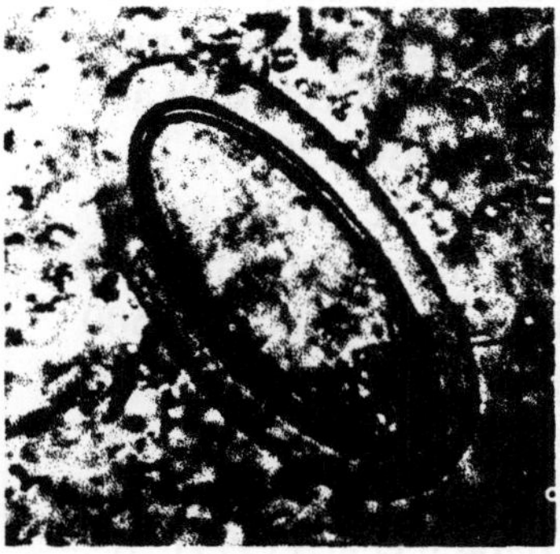

Figure 2. Egg of Moniliformis moniliformis *in a fecal smear.*

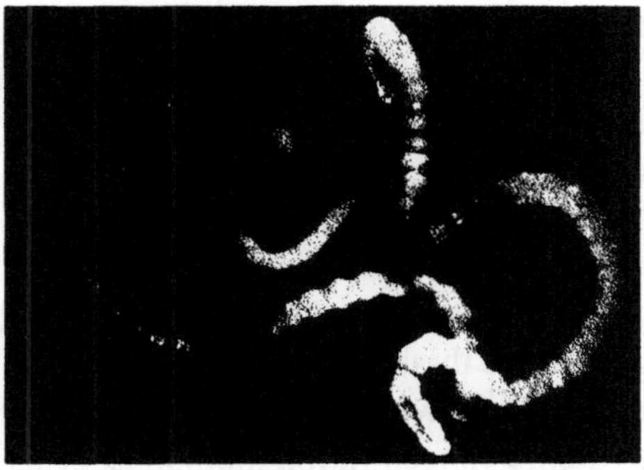

Figure 3. Moniliformis moniliformis *passed by a 15-month-old child in Florida. Courtesy Ron Neafie, AFIP.*

Treatment and Control Measures

All information regarding putative successful and unsuccessful treatment of acanthocephaliasis is anecdotal although several anthelmintics have been used in the treatment of acanthocephalan infections, including mebendazole (Goldsmid et al., 1974; Counselman et al., 1989; Prociv et al., 1990), thiabendazole (Moayedi, 1971), pyrantel pamoate (Goldsmid et al., 1974; Counselman et al., 1989; Prociv et al., 1990), ivermectin (Anoske, 2000), niclosamide (Sahba et al., 1970; Prociv et al., 1990) and Telmex (Sahba et al., 1970). Pyrantel pamoate (Counselman et al., 1989) and ivermectin (Anosike et al., 2000) have been reported to be the most successful. Both the *Red Book* (American Academy of Pediatrics, 2000) and *The Medical Letter* (2002) list pyrantel pamoate as the drug of choice for the treatment of moniliformiasis.

Studies are currently underway in the author's laboratory, using a rat model, to assess the effectiveness of various anthelmintics for the treatment of moniliformiasis. Both ivermectin and pyrantel pamoate were entirely *ineffective* in the treatment of moniliformiasis, even at very high dosages. No drug tested has affected complete clearance of infections in rats; however, thiabendazole shows promise since a single dose of 200 mg/kg (normal dosage for rats) reduced the mean worm burden by 40% after 4 days ($t = 2.04$; 9 d.f.; $P < 0.06$). Repeating the dose of thiabendazole at one week reduced the mean worm burden by 57% after 14 days ($t = 4.63$, 18 d.f.; $P < 0.001$). Based on available data, thiabendazole should tentatively be considered the drug of choice for the treatment of acanthocephaliasis at a pediatric dosage of 50 mg/kg/d in 2 doses (maximum of 3g/d) for 2 days as prescribed by *The Medical Letter* (2002) for the treatment of infection with the nematode *Strongyloides stercoralis*. This dosage of thiabendazole may be toxic (Medical Letter, 2002) and commonly causes side effects such as nausea, vomiting, malaise, and dizziness (Roberts and Janovy, 2000). If such side effects occur, the dosage should be reduced. If the patient is infected with *Ascaris lumbricoides*, the ascariasis should be treated first with albendazole (single dose of 400 mg), mebendazole (single dose of 500 mg), or pyrantel pamoate (single dose of 11 mg/kg with a maximum dose of 1 g) because thiabendazole may cause the ascarids to migrate. The patient should be followed closely with stool samples being examined weekly using the Ritchie concentration method as described by Markell et al. (1999). If eggs are observed, treatment should be repeated. The infection should not be considered "resolved" until no eggs are observed for 2 consecutive weeks *and* symptoms subside. Further studies are underway to assess the effectiveness of drugs with fewer side effects, such as mebendazole, for the treatment of moniliformiasis.

The only practical means of avoiding infection is to avoid the ingestion of raw cockroaches and beetles.

REFERENCES

Alonso A, Moreno-Ancillo A, Daschner A, López-Serrano MC. 1999. Dietary assessment in five cases of allergic reactions due to gastroallergic anisakiasis. *Allergy* 54: 517-520.

Al-Rawas AY, Mirza MY, Shafig MA. 1977. First finding of *Moniliformis moniliformis* (Bremser 1811) Travassos 1915 (Acanthocephala: Oligacanthorhynchidae) in Iraq from human child. *J Parasit* 63: 396-397.

Alvarez AM, Sanchez MC, Martin JM, Martin FS. 1990. Seroepidemiological studies on human pulmonary dirofilariasis in Spain. *Ann Trop Med Parasit* 84: 209-213.

American Academy of Pediatrics. 2000. Drugs for parasitic infections. In: Pickering LK (Ed.), *2000 Red Book: Report of the Committee on Infectious Diseases*, 25th ed. Elk Grove Village, IL: American Academy of Pediatrics. pp. 693-725.

Anderson EM, Lichtenfels JR. 2000. Anisakiasis. In: Meyers WM, Neafie RC, Marty AM, Wear DJ (Eds.), *Pathology of Infectious Diseases*, Volume I, *Helminthiases*. Washington, DC: Armed Forces Institute of Pathology. pp. 423-431.

Anderson RC, Morse PG, Scholten T. 1968. *Dirofilaria* (*Nochtiella*) infestation in a human in Ontario. *Canad Med Ass J* 98: 788.

Añíbarro B, Seoane FJ. 1998. Occupational conjunctivitis caused by sensitization to *Anisakis simplex*. *J Allergy Clin Immunol* 102: 331-332.

Anosike JC, Njoku AJ, Nwoke BEB, et al. 2000. Human infections with *Moniliformis moniliformis* (Bremser 1811) Travassos 1915 in south-eastern Nigeria. *Ann Trop Med Parasit* 94: 837-838.

Arh I. 1960. Fish tapeworm in Eskimos in the Port Harrison area, Canada. *Can J Pub Health* 51: 268-271.

Armentia A, Lombardero M, Callejo A, et al. 1998. Occupational asthma by *Anisakis simplex*. *J Allergy Clin Immunol* 102: 831-834.

Ash LR, Little MD. 1964. *Brugia beaveri* sp. n. (Nematoda: Filarioidea) from the raccoon (*Procyon lotor*) in Louisiana. *J Parasit* 50: 119-123.

Asimacopoulos PJ, Katras A, Christie B. 1992. Pulmonary dirofilariasis: The largest single-hospital experience. *Chest* 102: 851-855.

Baird JK, Alpert LI, Friedman R, et al. 1986. North American brugian filariasis: Report of nine infections in humans. *Am J Trop Med Hyg* 35: 1205-1209.

Baird JK, Klassen-Fischer MK, Neafie RC, Meyers WM. 2000. American brugian filariasis. In: Meyers WM, Neafie RC, Marty AM, Wear DJ (Eds.), *Pathology of Infectious Diseases*, Volume I, *Helminthiases*. Washington, DC: Armed Forces Institute of Pathology. pp. 319-328.

Beaver PC, Orihel TC. 1965. Human infection with filariae of animals in the United States. *Am J Trop Med Hyg* 14: 1010-1029.

Beaver PC, Wolfson JS, Waldron MA, et al. 1987. *Dirofilaria ursi*-like parasites acquired by humans in the northern United States and Canada: Report of two cases and a brief review. *Am J Trop Med Hyg* 37: 357-362.

Beaver PC, Wong MM. 1988. *Brugia* sp. from a domestic cat in California. *Proc Helminthol Soc Was* 55: 111-113.

Beck JW. 1959. Report of a possible human infection with the acanthocephalan *Moniliformis moniliformis* (syn. *M. dubius*). *J Parasil* 45: 510.

Berk SL, Verghese A, Alvarez S, et al. 1987. Clinical and epidemiologic features of strongylodiasis: A prospective study in rural Tennessee. *Arch Intern Med* 147: 1257-1261.

Carretero Anibarro P, Blanco Carmona J, Garcia Gonzalez F, et al. 1997. Protein contact dermatitis caused by *Anisakis*. *Contact Dermatitis* 37: 247.

CDC. 1981. Cutaneous larva migrans in American tourists-Martinique and Mexico. *MMWR* 30: 308,313.

CDC. 1982. Cercarial dermatitis among bathers in California; Katayama syndrome among travelers to Ethiopia. *MMWR* 31: 435-438.

CDC. 1992. Cercarial dermatitis outbreak at a state park—Delaware, 1991. *MMWR* 41: 225-228.
Collins BM, Jones AC, Jimenez F. 1993. *Dirofilaria tenuis* infection of the oral mucosa and cheek. *J Oral Maxillofac Surg* 51: 1037-1040.
Conly JM, Sekla LH, Low DE. 1984. Dirofilariasis presenting as a breast lump. *Can Med Assoc J* 130: 1575-1576.
Corman LC. 1987. Acute arthritis occurring in association with subcutaneoius *Dirofilaria tenuis* infection. *Arthritis Rheum* 1431-1434.
Counselman K, Field C, Grady L, et al. 1989. *Moniliformis moniliformis* from a child in Florida. *Am J Trop Med Hyg* 41: 88-90.
Croese J, Loukas A, Opdebeeck J, et al. 1994. Human enteric infection with canine hookworms. *Ann Intern Med* 120: 369-374.
Crompton DWT. 1999. How much human helminthiasis is there in the world? *J Parasit* 85: 397-403.
Cuende E, Audicana MT, Garcia M, Anda M. 1998. Rheumatic manifestations in the course of anaphylaxis caused by *Anisakis simplex*. *Clin Exp Rheumatol* 16: 303-304.
Dalton JP (Ed.). 1999. *Fasciolosis*. Wallingford, UK: CABI Publishing. 562 pp.
Daly JJ. 1982. Sparganosis. In: Arambulo P (Ed.), *CRC Handbook Series in Zoonoses Section C: Parasitic Zoonoses,* Volume I, Part 2. Boca Raton, FL: CRC Press, Inc. pp. 293-312.
Daschner A, Alonso-Gómez A, Cabañas R, et al. 2000. Gastroallergic anisakiasis: Boderline between food allergy and parasitic diseaes—clinical and allergologic evaluation of 20 patients with confirmed acute parasitism by *Anisakis simplex*. *J Allergy Clin Immunol* 105: 176-181.
Davies P, Cunningham TJ, Gould RG, Horton J. 1973. *Dirofilaria* infection: Occurrence in northeastern New York. *New York State J Med* 298: 1999-2001.
Deardorff TL, Kayes SG, Fukumura T. 1991. Human anisakiasis transmitted by marine food products. *Hawaii Med J* 50: 9-16.
Del Pozo MD, Audícana M, Diez JM, et al. 1997. *Anisakis simplex*, a relevant etiologic factor in acute urticaria. *Allergy* 52: 576-579.
Despommier DD, Gwadz RW, Hotez PJ, Knirsch CA. 2000. *Parasitic Diseases*, 8th ed. New York, NY: Apple Trees Productions, LLC. 345 pp.
Dingley D, Beaver PC. 1985. *Macracanthorhynchus ingens* from a child in Texas. *Am J Trop Med Hyg* 34: 918-920.
Eastburn RL, Fritsche TR, Terhune CA Jr. 1987. Human intestinal infection with *Nanohyetus salmincola* from salmonid fishes. *Am J Trop Med Hyg* 36: 586-591.
Eberhard ML. 1984. *Brugia lepori* sp. n. (Filarioidea: Oncocercidae) from rabbits (*Sylvilagus aquaticus, S. floridanus*) in Louisiana. *J Parasit* 70: 576-579.
Eberhard ML, Telford SR III, Spielman A. 1991. A *Brugia* species infecting rabbits in the northeastern United States. *J Parasit* 77: 796-798.
Eberhard ML, DeMeester LJ, Martin BW, Lammie PJ. 1993. Zoonotic *Brugia* infection in western Michigan. *Am J Surg Pathol* 17: 1058-1061.
Elenitoba-Johnson KSJ, Eberhard ML, Dauphinais RM, et al. 1996. Zoonotic brugian lymphadenitis: An unusual case with florid moncytoid B-cell proliferation. *Am J Clin Pathol* 105: 384-287.
Feldman GJ, Parker HW. 1992. Visceral larva migrans associated with hypereosinophilic syndrome and the onset of severe asthma. *Ann Internal Med* 116: 838-840.
Genta RM. 1989. Global prevalence of strongylodiasis: Critical review with epidemiological insights in the prevention of disseminated disease. *Rev Infect Dis* 2: 755-767.
Georgi JR, Sprinkle CL. 1974. A case of human strongyloidosis apparently contracted from asymptomatic colony dogs. *Am J Trop Med Hyg* 23: 899-901.
Goldsmid JM, Smith ME, Fleming F. 1974. Human infection with *Moniliformis* sp. in Rhodesia. *Ann Trop Med Parasit* 68: 363-364.

Grassi B, Calandruccio S. 1888. Ueber einen Echinorhynchus, welcher auch in Menschen parasitirt und dessen Zwischenwirt ein Blaps ist. *Zentralbl Bakt Parasitenkd Orig* 3: 521-525.

Greene BM, Otto GF, Greenough WB III. 1978. Circulating non-human microfilaria in a patient with systemic lupus erythematosus. *Am J Trop Med Hyg* 27: 905-909.

Grove DI (Ed.). 1989. *Strongyloidiasis: A Major Roundworm Infection of Man*. LONDON: Taylor and Francis. 336 pp.

Hendrix CM. 1995. Helminth infections of the feline small and large intestines: Diagnosis and treatment. *Vet Med* May 1995: 456-472.

Hoeffler DF. 1974. Cercarial dermatitis: Its etiology, epidemiology, and clinical aspects. *Arch Environ Health* 29: 225-229.

Isaza R, Courtney CH. 1988. Possible association between *Dirofilaria tenuis* infections in humans and its prevalence in raccoons in Florida. *J Parasit* 74: 189-190.

Jackson GJ. 1975. The "new disease" status of human anisakiasis and North American cases: A review. *J Milk Food Technol* 38: 769-773.

Jarrett E, Mackenzie S, Bennich H. 1980. Parasite-induced 'nonspecific' IgE does not protect against allergic reactions. *Nature* 283: 302-304.

Jelinek T, Maiwald H, Nothdurft HD, Löscher T. 1994. Cutaneous larva migrans in travelers: Synopsis of histories, symptoms, and treatment of 98 patients. *CID* 19: 1062-1066.

Johansson E, Aponno M, Lundberg M, van Hage-Hamsten M. 2001. Allergenic cross-reactivity between the nematode *Anisakis simplex* and the dust mites *Acarus siro, Lepidoglyphus destructor, Tyrophagus putrescentiae,* and *Dermatophagoides pteronyssinus*. *Allergy* 56: 660-666.

Keeling NG, Roelke ME, Forrester DJ. 1993. Subcutaneous helminthes of the raccoon (*Procyon lotor*) in southern Florida. *J Helmthol Soc Wash* 60: 115-117.

Kliks M, Tantachamrun T, Chaiyaporn V. 1974. Human infection by an acanthocephalan *Macracanthorhynchus hirudinaceus* in Thailand: New light on a previous case. *Southeast Asian J Trop Med Pub Heath* 5: 303-309.

Levinson ED, Ziter FM Jr, Westcott JL. 1979. Pulmonary lesions due to *Dirofilaria immitis* (dog heartworm). *Radiology* 131: 305-307.

Little MD, Halsey NA, Cline BL, Katz SP. 1983. *Ancylostoma* larva in muscle fiber of man following cutaneous larva migrans. *Am J Trop Med Hyg* 32: 1285-1288.

Loken BR, Spencer CN, Granath WO Jr. 1995. Prevalence and transmission of cercariae causing schistosome dermatitis in Flathead Lake, Montana. *J Parasit* 81: 646-649.

Lorenzo S, Iglesias R, Leiro J, et al. 2000. Usefulness of currently available methods for the diagnosis of *Anisakis simplex* allergy. *Allergy* 55: 627-633.

MacLean JD, Arthur JR, Ward BJ, et al. 1996. Common-source outbreak of acute infection due to the North American liver fluke *Metorchis conjunctus*. *Lancet* 347: 154-158.

Markell EK, John DT, Krotski WA. 1999. *Markell and Voge's Medical Parasitology*, 8th ed. Philadelphia, PA: W. B. Saunders Co. 501 pp.

Marquardt WC, Demaree RS, Grieve RB. 2000. *Parasite and Vector Biology*, 2nd ed. San Diego CA: Harcourt Academic Press. 702 pp.

Marty AM, Neafie RC. 2000a. Diphyllobothriasis and sparganosis. In: Meyers WM, Neafie RC, Marty AM, Wear DJ (Eds.), *Pathology of Infectious Diseases*, Volume I, *Helminthiases*. Washington, DC: Armed Forces Institute of Pathology. pp. 165-183.

Marty AM, Neafie RC. 2000b. Dirofilariasis. In: Meyers WM, Neafie RC, Marty AM, Wear DJ (Eds.), *Pathology of Infectious Diseases*, Volume I, *Helminthiases*. Washington, DC: Armed Forces Institute of Pathology. pp. 275-285.

Mas-Coma S, Bargues MD, Esteban JG. 1999. Human fasciolosis. In: Dalton JP (Ed.). *Fasciolosis*. Wallingford, UK: CABI Publishing. pp. 411-434.

The Medical Letter. 2002. Abramowicz M (Ed.). New Rochelle, NY: The Medical Letter, Inc. 12 pp.

McKerrow TL, Sakanari J, Deardorff TL. 1988. Anisakiasis: Revenge of the sushi parasite. *N Engl J Med* 319: 1228-1229.
Merck and Co., Inc. 1998. Zoonoses. In: Aiello SE (Ed.), *The Merck Veterinary Manual*, 8th ed. Whitehouse Station, NJ: Merck and Co., Inc. pp. 2160-2185.
Millemann RE, Knapp SE. 1970. Biology of *Nanophyetus salmincola* and "salmon poisoning" disease. *Adv Parasit* 8: 1-41.
Moayedi B, Izadi M, Maleki M, Ghadirian E. 1971. Human infection with *Moniliformis moniliformis* (Bremser, 1811) Travassos, 1915 (syn. *Moniliformis dubius*): Report of a case in Isfahan, Iran. *Am J Trop Med Hyg* 20: 445-448.
Moneo I, Caballero ML, Gómez F, et al. 2000. Isolation and characterization of a major allergen from the fish parasite *Anisakis simplex*. *J Allergy Clin Immunol* 106: 177-182.
Moreno-Ancillo A, Caballero MT, Cabañas R, et al. 1997. Allergic reactions to *anisakis simplex* parasitizing seafood. *Ann Allergy Asthma Immunol* 79: 246-250.
Muhleisen JP. 1953. Demonstration of pulmonary migration of the causative organism of creeping eruption. *Ann Intern Med* 38: 595-600.
Neafie RC, Marty AM. 2000. Acanthocephaliasis. In: Meyers WM, Neafie RC, Marty AM, Wear DJ (Eds.), *Pathology of Infectious Diseases*, Volume I, *Helminthiases*. Washington, DC: Armed Forces Institute of Pathology. pp. 519-529.
Nelson MJ, Nickol BB. 1986. Survival of *Macracanthorhynchus ingens* in swine and histopathology of infection in swine and raccoons. *J Parasit* 72: 306-314.
Norton RA, Monroe L. 1961. Infection by *Fasciola hepatica* acquired in California. *Gastroenterology* 41: 46-48.
Orihel TC, Beaver PC. 1989. Zoonotic *Brugia* infections in North and South America. *Am J Trop Med Hyg* 40: 638-647.
Pascual CY, Crespo JF, San Martin S, et al. 1997. Cross-reactivity between IgE-binding proteins from *Anisakis*, German cockroach, and chironomids. *Allergy* 52: 514-520.
Pawlowski ZS. 1989. Epidemiology, prevention and control. In: Grove DI (Ed.), *Strongyloidiasis: A Major Roundworm Infection of Man*. London: Taylor and Francis. pp. 233-249.
Payan HM. 1978. Human infection with *Dirofilaria*. *Arch Dermatol* 114: 593-594.
Perez Sanchez R, Gomez Bautista M, Encinaas Grandes A. 1989. Canine filariasis in Salamanca (Northwest Spain). *Ann Trop Med Parasit* 83: 143-150.
Pradatsundarasar A, Pechranond K. 1965. Human infection with the acanthocephalan *Macracanthorhynchus hirudinaceus* in Bangkok: Report of a case. *AmJ Trop Med Hyg* 14: 774-776.
Prociv P, Walker J, Crompton LJ, Tristram SG. 1990. First record of human acanthocephalan infections in Australia. *Med J Aust* 152: 215-216.
Pung OJ, Davis PH, Richardson DJ. 1996. Filariae of raccoons from southeast Georgia. *J Parasit* 82: 849-851.
Radomyos P, Chobchuanchom A, Tungtrongchitr A. 1989. Intestinal perforation due to *Macracanthorhynchus hirudinaceus* infection in Thailand. *Trop Med Parasit* 40: 476-477.
Richardson DJ, Barnawell EB. 1995. Histopathology of *Oligacanthorhynchus tortuosa* (Oligacanthorhynchidae) infection in the Virginia opossum (*Didelphis virginiana*). *J Helminthol Soc Was* 62: 253-256.
Roberts LS, Janovy J Jr. 2000. *Gerald D. Schmidt and Larry Robert's Foundations of Parasitology*, 6th ed. Boston, MA: McGraw Hill Co., Inc., 670 p.
Roy BT, Chirurgi VA, Theis JH. 1993. Pulmonary dirofilariasis in California. *West J Med* 58: 74-76.
Sahba GH, Arfaa F, Rastegar M. 1970. Human infection with *Moniliformis dubius* (Acanthocephala) (Meyer, 1932). (syn. *M. moniliformis*, (Bremser, 1811) (Travassos, 1915) in Iran. *Trans Royal Soc Trop Med Hyg* 64: 284-286.

Sastre J, Lluch- Bernal M, Quirce S, et al. 2000. A double-blind, placebo-controlled oral challenge study with lyophilized larvae and antigen of the fish parasite, *Anisakis simplex*. *Allergy* 55: 560-564.
Schantz PM. 1989. The dangers of eating raw fish. *N Engl J Med* 320: 1143-1145.
Schmidt GD. 1972. Acanthocephala of captive primates. In: T-W-Fiennes RN (Ed.), *Pathology of Simian Primates*, Part II. Basel: S. Karger. pp. 144-156.
Schmidt GD. 1985. Development and life cycles. In: Crompton DWT, Nickol BB (Eds.), *Biology of the Acanthocephala*. Cambridge, UK: Cambridge Univ Press. pp. 273-305.
Sharghi N, Schantz PM, Caramico L, et al. 2001. Environmental exposure to *Toxocara* as a possible risk factor for asthma: A clinic-based case-control study. *CID* 32: e111-e116.
Simmons CF, Winter HS, Berde C. 1984. Zoonotic filariasis with lemphedema in an immunodeficient infant. *N Engl J Med* 310: 1243-1245.
Smith JW, Wootten R. 1970. *Anisakis* and anisakiasis. *Adv in Parasit* 16: 93-163
Stoll NR. 1947. This wormy world. *J Parasit* 33: 1-18.
Sulica VI, Berberian B, Kao GF. 1988. Histopathologic findings of cutaneous larva migrans. *J Cutan Pathol* 15: 346.
Taraschewski H, Sagani C, Mehlhorn H. 1989. Ultrastructural study of the host-parasite interface of *Moniliformis moniliformis* (Archiacanthocephala) in laboratory-infected rat. *J Parasit* 75: 288-296.
Theis JH, Gilson A, Simon GE, et al. 2001. Case report: Unusual location of *Dirofilaria immitis* in a 28-year-old man necessitates orchiectomy. *Am J Trop Med Hyg* 64: 317-322.
Tullis DCH. 1970. Bronchial asthma associated with intestinal parasites. *N Engl J Med* 282: 370-372.
Turner KJ. 1978. The conflicting role of parasitic infections in modulating the prevalence of asthma. *Papua N Guinea Med J* 21: 86-104.
Vienn AL, Adoma C, Ward Y. 2000. *Macracanthorhynchus ingens* worm: An uncommon parasite. *Southwest Assoc Clin Micro 2000*: p. 100.

CRYPTOSPORIDIOSIS

Cynthia L. Chappell and Pablo C. Okhuysen

University of Texas, School of Public Health, Houston, Texas

Etiological Agent: *Cryptosporidium parvum*; protozoan of the small intestinal lining
Epidemiology: *Cryptosporidium parvum* consists of two major genotypes with different transmission cycles. Both genotypes cause diarrheal illness in immunocompetent and immunocompromised individuals. Other *Cryptosporidium* species are associated with a small percentage of human infections. *Cryptosporidium* seroprevalence ranges from about 25-35 percent in the U.S., but may be higher in certain populations.
Clinical Manifestations: Clinical manifestations in humans may range from asymptomatic to a profuse, watery diarrhea. Moderate to severe infections are characterized by diarrhea and abdominal pain and cramping. Nausea may also occur but is not usually accompanied by vomiting.
Complications: Complications occur only in malnourished children or immunocompromised persons who may experience a chronic and progressive diarrhea. In these individuals, the infection may be life-threatening. Persons with intact immune systems have a self-limited infection.
Mode of Transmission: Infection may be acquired through the fecal-oral route from an infected person or animal or ingestion of contaminated drinking or recreational water.
Diagnosis: Microscopic detection of fecal oocysts by Ziehl-Neelsen stain or monoclonal antibody-based ELISAs or IFA. Experimental techniques include detection of fluorescent-labelled oocysts by flow cytometry and PCR detection of parasite DNA using genus or species-specific primers.
Drug(s) of Choice: No curative treatment is available. The combination of paromomycin and azithromycin may be helpful in lessening the symptoms and decreasing oocyst shedding.
Reservoir Hosts: 179 mammalian species have been documented, including common infections of neonatal calves, lambs, goats and other herbivores.
Control Measures: Use of good hygienic practices, such as thorough hand washing after bowel movements. Animal control in water catchment areas. Management of animal and human waste water. Maintenance of high quality water treatment facilities to lessen the number of viable oocysts in the drinking water supply.

Cryptosporidium parvum was first identified by Edward Tyzzer early in the 20th century (Tyzzer, 1912). However, it was not until 1976 that the first human case of cryptosporidiosis was diagnosed in a 3-year-old child living on a Tennessee farm (Nime et al., 1976). This paper was quickly followed by a case report in an immunosuppressed individual (Meisel et al., 1976). The interest in the parasite, however, was increased dramatically when it was recognized in the early 1980's as a serious opportunistic infection of AIDS patients (Anon., 1982). During this same time, outbreaks of cryptosporidiosis were being recognized in a variety of situations, including contamination of drinking water (D'Antonio et al., 1985; Hayes et al., 1989), swimming pools (Sorville, 1990), and day care centers (Soave and Armstrong, 1986). The largest known outbreak of waterborne disease occurred in Milwaukee in 1993, where approximately 400,000 persons experienced a diarrheal illness (MacKenzie et al., 1994). The causative agent was confirmed as *Cryptosporidium parvum* which was distributed from one of the water treatment plants in the city. Approximately 100 deaths in AIDS patients were attributed to the infection.

Etiology/Natural History

Cryptosporidium is a protozoan parasite that infects the epithelial lining of mucosal tissues and undergoes both asexual and sexual replication within the same host (Figure 1). Heaviest infection is thought to involve the enterocytes of the ileum, but can spread to contiguous epithelial cells if

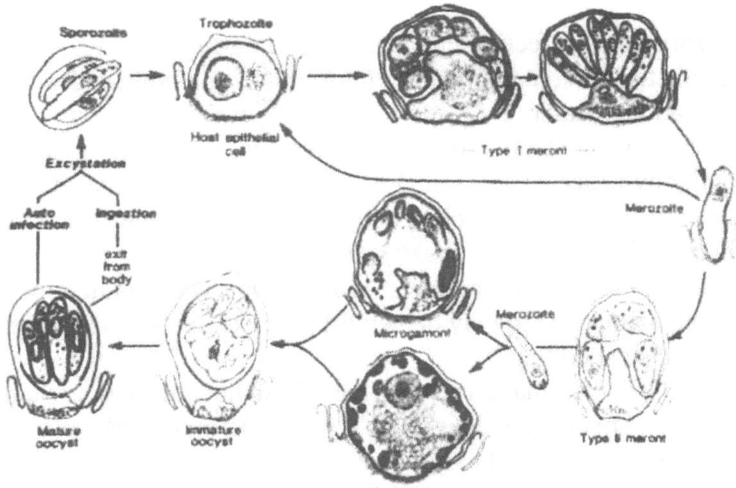

Figure 1. The life cycle of Cryptosporidium. *(From Fayer, R., ed., in Cryptosporidium and Cryptosporidiosis, CRC Press, Boca Raton, FL, 1997. With permission.)*

unchecked by an intact immune system. The oocyst consists of a tough, protective covering for four sporozoites. The oocyst wall is environmentally resistant and can withstand most detergents and disinfectants (Figure 2). Once ingested, the oocyst is acted upon by gastric acid, digestive enzymes, and bile that induce excystation in the small bowel. Released sporozoites are motile and quickly search out enterocytes, which they attach to and invade. Although the parasite develops intracellularly, the cellular location is just below the plasma membrane, which along with parasite-derived membrane

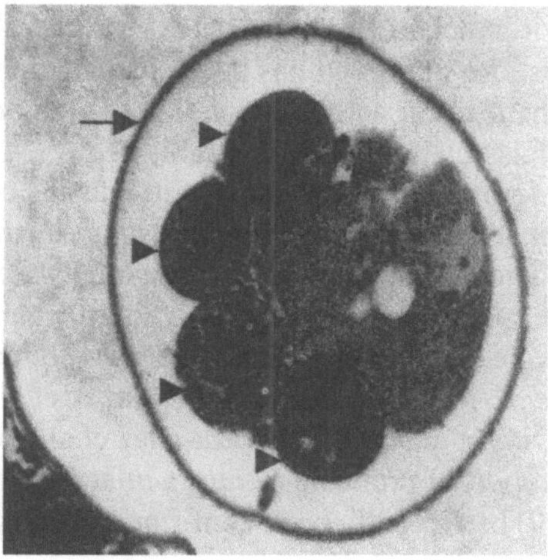

Figure 2. Cross-section of a Cryptosporidium parvum *oocyst. Note the thick dense wall (arrow) surrounding four sporozoites (arrowheads). Photo courtesy of Dr. Rebecca Langer, University of Texas School of Public Health, Houston, Texas.*

forms the parasitophorous vacuole. Thus, the organism does not enter the cytoplasm of the host cell nor does it gain access to submucosal tissues. The base of the parasitophorous vacuole is characterized by a convoluted electron-dense membrane called the feeder organelle, that is thought to regulate the passage of host cell constituents needed by the parasite for its growth and development. The sporozoite undergoes asexual division within the parasitophorous vacuole to form 6 or 8 Type I merozoites. Merozoites erupt from the vacuole and search out neighboring epithelial cells to invade and continue the replicative cycle. Other coccidians are known to have limited asexual cycles; however, this may not be the case for *Cryptosporidium*, which can cause chronic infections. After the first replication cycle, a portion of the merozoites develop into Type II merozoites. Type II merozoites form the sexual stage of the life cycle developing into either a single macrogamont or multiple microgametes. The microgametes are released from the host cell and seek out a cell containing a macrogamont by a mechanism that is as yet unknown. The successful microgamont penetrates the vacuole containing the macrogamont and fuses with it, giving rise to a zygote. The zygote continues its development within the parasitophorous vacuole, and the oocyst wall begins to form around it. The oocyst and its contents are released from the vacuole and are swept out of the GI tract along with the feces. Sporulation takes place while the oocyst is still within the host, so that when the oocyst is released it is immediately infectious for other susceptible hosts. There appear to be two kinds of oocysts that are released into the intestinal lumen, the thick-walled oocyst that can withstand environmental conditions and the thin-walled oocyst, which is thought to be more fragile. It has been postulated that thin-walled oocysts may be a mechanism for autoinfection, but the importance of this putative route of transmission has not been established.

The most common form of cryptosporidiosis in humans comes from infection by *C. parvum*, although recent evidence suggests that both immunocompromised and immunocompetent individuals are susceptible to other *Cryptosporidium* species once thought only to infect animal species. The "non-*parvum*" species that have been associated with human infections include *C. meleagridis*, *C. felis*, and *C. canis* (Pieniazek et al., 1999; Morgan et al., 2000; Pedraza-Diaz et al., 2000; Pedraza-Diaz et al., 2001; Xiao et al., 2001). The first case report of probable *C. muris* infection in two young girls (Katsumata et al., 2000) has recently been confirmed in an AIDS patient (Gatei et al., 2002).

C. parvum is now recognized to include two distinct genotypes, each with a specific transmission cycle. Genotype 1 (human strain) oocysts are transmitted directly or indirectly from one infected person to another. Cross-species infection studies have not been successful in transmitting this genotype to herbivores or other mammals. The only successful replication of genotype 1 oocysts in a laboratory animal involves piglets raised in a gnotobiotic environment and infected as neonates (Widmer et al., 2000). In contrast, genotype 2 (bovine species) oocysts may be transmitted between animals and humans or between animals, such as in a herd. Thus, humans may become infected and ill from either of these two *C. parvum* genotypes, while animal infections are limited to Type 2 oocysts under normal conditions. There is accumulating evidence that these genotypes may, in fact, constitute two distinct *Cryptosporidium* species. Molecular evidence and the lack of recombination between the strains strongly suggests that this is so and has provided the taxonomists with much to consider.

Epidemiology

Evidence for widespread exposure to *Cryptosporidium* is seen in the large number of persons who have evidence of parasite-specific serum antibodies. In the U.S., the percentage of seropositive individuals varies by population.

In most urban areas, seropositivity is in the range of 25-35 percent but may be higher in rural populations, particularly in those settings where there is animal contact or where sanitation is poor. In the latter cases, seropositive individuals may comprise 80 percent or more of the population (Leach et al., 2000). In developing countries, seroprevalences are significantly higher, especially in rural populations, and may reach 60 percent or greater (reviewed in Ungar, 1990). Travelers exposed to the local water supply in such areas may become infected (Thielman and Guerrant, 1998).

In general terms, children (reviewed in Casemore et al., 1997) and elderly persons (Bannister and Mountford, 1989) appear to be more susceptible to infection and/or develop a more severe illness. High rates of *Cryptosporidium* infection may be found in day care centers with typical prevalences in the range of 30-40 percent (reviewed in Ungar, 1990; Cordell and Addiss, 1994). Many of these infections are asymptomatic.

In HIV-infected individuals, prevalence of infection and illness is related to the $CD4^+$ lymphocyte count. Chronic cryptosporidiosis is typically seen in AIDS patients with $CD4^+$ counts below $200mm^3$. In the days prior to HAART, the prevalence of *Cryptosporidium* infection in AIDS patients with diarrhea ranged from 15-33 percent in developed countries (Smith et al., 1988; Rene et al., 1989; Brandonisio et al., 1993). Higher prevalences were reported in AIDS patients in developing countries. Type 1 *Cryptosporidium* infections appear to be predominant in the HIV population (Widmer et al., 1998), which may suggest a greater role from direct person-to-person transmission rather than exposure via contamination of the drinking water supply. HIV-infected children are also susceptible to chronic cryptosporidiosis, although the prevalence of infection (about 8 percent) appears to be lower than that of adults (Navin and Hardy, 1987; Guarino et al., 1997). Other immunocompromising conditions resulting in low CD4+ cell counts, such as cancer and its treatments, may be associated with chronic cryptosporidiosis and basically mimic the presentation seen in HIV infections (Tanyuksel et al., 1995; Gentile et al., 1991).

Early indications are that *Cryptosporidium* species other than *C. parvum* account for a low proportion (3.5-5 percent) of all diarrheal cases (Pedraza-Diaz et al., 2000; Pedraza-Diaz et al., 2001), but up to 36 percent of HIV-associated cryptosporidiosis (Morgan et al., 2000). It should be noted, however, that these prevalence estimates may be revised as the molecular probes to detect various *Cryptosporidium* species are refined and as larger populations are tested.

Cryptosporidium has been associated with a number of diarrheal outbreaks in communities throughout the U.S. and other developed countries. Many of these outbreaks have been linked to the public water supply or recreational water (rivers, lakes and water parks) and presumably contribute greatly to the high seroprevalence. Earlier studies of environmental waters indicated that the parasite is widespread throughout the U.S. and can also be isolated from wells (LeChevallier et al., 1991). However, the proportion of diarrheal cases caused by waterborne cryptosporidiosis is unclear, especially in developed areas. Two recent studies in Australian cities indicated that little to no *Cryptosporidium* infections were contracted from the drinking water supply, even though one of the cities had poor quality source water (Hellard et al., 2001). Further, the major risk factors associated with cryptosporidiosis in the studies were swimming in public pools and contact with infected persons. If this scenario holds true in other developed countries, control measures for recreational parks and institutional settings, such as day care centers may need to be re-evaluated.

Waterborne outbreaks of cryptosporidiosis have been associated with both Type 1 and Type 2 oocysts. The *Cryptosporidium* genotype responsible for such outbreaks varies by geographic location and season (Dietz et al., 2000; McLauchlin et al., 2000). In a recent study in the UK, 13 outbreaks associated with drinking or recreational waters found that 7 were associated with genotype 1, 4 were associated with genotype 2, and 2 were associated with both genotypes (McLauchlin et al., 2000). They further found that family and day care center outbreaks were equally divided between the two genotypes. In an earlier UK study, sporadic disease was primarily (59 percent) associated with genotype 1, while genotype 2 and mixed genotype infections accounted for 35 percent and 6 percent, respectively (Patel et al., 1998). One may speculate that Type 1 outbreaks may arise from person-to-person transmission or source water contamination with human sewage, while Type 2 outbreaks may come about, in part, from surface water runoff of pasturelands. Contamination of source water with either oocyst type poses a public health threat since the parasite is difficult to remove during water treatment processing and is incompletely inactivated by the amount of chlorine added to drinking water. New molecular epidemiological methods now under development should assist in tracking particular strains in communities and in outbreaks. Further, the water quality industry is active in developing new testing strategies for monitoring the infectivity of environmental oocysts that have found their way into source waters (Rochelle and De Leon, 2001).

Foodborne outbreaks of *Cryptosporidium* have been documented, but appear to be rare. Contamination of food during preparation is the most commonly reported scenario (Anonymous, 1998;Quiroz et al., 2000). However, other instances of contamination have been documented. One instance involving 160 diarrheal cases was associated with homemade apple cider prepared for a school fair (Millard et al., 1994). In this particular situation, apples had been collected from trees in a field where cattle grazed, and some of the apples had been picked up off the ground. The cider and apple press from which it was prepared were found to contain oocysts. In another case, the parasite was transmitted in food prepared for a small social event by a caterer who also managed a small day care center in her home (Besser-Wiek et al., 1996). *Cryptosporidium* was confirmed in the stools of attendees, although no oocysts were recovered from the food. Contamination presumably had come from caring for one or more infected children in the day care group.

Pathogenesis

A number of pathogenic mechanisms for *Cryptosporidium* have been identified and recently reviewed (Okhuysen and Chappell, in press). Most of these mechanisms are associated with the initial attachment and invasion of the parasite into host enterocytes. Others appear to be involved in the disruption of host membranes or the elimination of host cells via apoptosis. Little is known about factors that control the replication and development of the parasite within the parasitophorous vacuole. The factors that are responsible for causing diarrhea have been a matter of much speculation. Although the diarrhea is secretory in nature, no toxin has been identified despite serious and thorough investigations. As more is known about the cytokine response to the organism, the cascade of events and their associated molecules leading to diarrhea may become more apparent. At least one investigator has suggested that cryptosporidial diarrhea may be mediated by TNFα via prostaglandins (Kandil et al., 1994). However, in recent volunteer studies TNFα was not limited to subjects experiencing a diarrheal illness. Additional work will undoubtedly be necessary to understand the sequence of events leading to diarrhea in some individuals, while *Cryptosporidium* infections in others remain asymptomatic.

Clinical manifestations and complications

Diarrhea and abdominal pain are characteristic features of cryptosporidiosis and are often accompanied by malaise, flatulence, bloating, fecal urgency, and tenesmus. Diarrhea may be mild to severe with profuse, watery stools. Other systemic symptoms, such as nausea, fever, vomiting, and myalgia may also occur, but are more rare and when they do occur are not the prominent symptoms. In the immunocompetent host, these symptoms typically resolve within one to two weeks, but diarrhea of one month has been occasionally seen. Thus, the infection in these individuals is self-limited.

In general populations and outbreak situations, cryptosporidiosis has been observed to have an incubation period of approximately 6 to 10 days after exposure. The typical diarrheal illness lasts for 2 to 26 days (mean 12 days) before it slowly resolves. The variable clinical course in the general population reflects such factors as general health, immune competence, potential concurrent enteric bacterial or viral infections, and the virulence of the infecting *Cryptosporidium* strain. In some individuals, symptoms may be intermittent with one or two recurrences a few days apart. It is likely that these recurrences arise during the course of a single infection, but this is difficult to demonstrate directly.

Healthy, adult volunteers each experimentally-challenged with one of four strains of *Cryptosporidium* parvum (Type 2) oocysts have been very valuable in describing the natural history of the infection and illness. A total of 85 subjects with no prior evidence of *Cryptosporidium* infection (i.e. IgG negative in ELISA) were challenged with one of four *C. parvum* isolates. In these studies, persons with oocyst-confirmed infections experienced a variety of outcomes, even in those receiving the same parasite strain. Asymptomatic infections were observed in 12.9 percent of all participants, but were strongly associated with only one of the isolates. The illness attack rate also varied from 52 percent to 86 percent depending upon the *Cryptosporidium* strain used (DuPont et al., 1995; Okhuysen et al., 1999; Okhuysen et al., in press) and the volunteer's previous exposure to the parasite (Chappell et al., 1999). For all four of the isolates taken together, these studies indicate that the mean incubation time among strains ranged from 4 to 7 days post challenge, and diarrhea lasted for 4 to 7 days. Illness resolved in 83 percent of volunteers within 14 days and 100 percent by 21 days post onset. Likewise, oocyst shedding cleared in 73 percent of volunteers within 14 days of onset, and 93 percent cleared by 21 days. The last 7 percent of volunteers cleared the infection within 28 days. Overall, recurrent illness was observed in 12 percent of all volunteers with diarrhea and were not observed in any of the 27 persons with previous exposure to the parasite. Mean stool frequencies per diarrheal episode ranged from 8 to 19. One of the strains, however, was associated with the most severe outcomes. In comparison to the other strains, the Moredun isolate had a shorter incubation time (mean=4.2 days post challenge), a longer duration of illness (mean=6.9 days) and a greater stool weight per episode (mean=1836 grams).

Oocyst shedding typically began 6 to 8 days after exposure and in 33 percent of volunteers continued for a mean of 8 days after resolution of the illness. For all isolates, the median number of *Cryptosporidium* oocysts shed over the course of the infection was in the range of 1.2-5.5 X 10^6 with only 11 percent of individuals (all strains) excreting 10^8 or more oocysts. These numbers are in contrast to infections in neonatal calves or AIDS patients, where shedding of 10^9 to 10^{10} oocysts is common.

Although the general clinical features of the illness are the same as in immunocompetent individuals, the immunocompromised patient may have a chronic and progressive disease. In outbreak situations, this population has been shown to be particularly susceptible to the infection and suffer the severest consequences (MacKenzie et al., 1994). In addition to those with HIV infections, an increased risk of chronic cryptosporidiosis has been associated with malignancies (Gentile et al., 1991; Tanyuksel et al., 1995), renal transplant and dialysis (Chieffi et al., 1998), and other conditions in which the immune system is adversely affected as a result of disease or suppressive drugs. However, in the U.S. mortality associated with this parasite has been in the AIDS population. Before the development of highly active antiretroviral therapy (HAART), it was estimated that 10-15 percent of AIDS patients in the U.S. would contract *Cryptosporidium* at some point in the course of their HIV infection (Laughon et al., 1988). Patients with $CD4^+$ T cells below 180 mm^3 were at a greatly increased risk of serious and chronic cryptosporidiosis. In HIV-infected persons, *Cryptosporidium* spreads from the nidus of infection into the adjacent mucosa and may extend from the ileum, where the most intense infection is thought to occur, to the colon and upstream into the jejunum and duodenum. The bile ducts and gall bladder epithelium often become involved. The organism has also been documented in the stomach (Rivasi et al., 1999), respiratory tree (Clavel et al., 1996) and even in the middle ear and sinuses (Dunand et al., 1997), presumably from spread during vomiting and aspiration. The overall incidence in cryptosporidiosis in the HIV population has changed dramatically with HAART since the post treatment increase in $CD4^+$ lymphocyte counts reconstitutes the immune system and results in self-limited *Cryptosporidium* infection (Carr et al., 1998).

AIDS patients are not the only population that is especially sensitive to this organism. Persons with primary immunodeficiencies (Kocoshis et al., 1984; Jacyna et al., 1990; Gomez-Morales et al., 1996) or those receiving cancer chemotherapy or other immunosuppressive drugs may be at increased risk for chronic *Cryptosporidium* infection. When this latter situation occurs, the patient must either endure the symptoms or the immunosuppressive treatments must be discontinued, when possible, until the immune functions are restored. In underdeveloped countries the situation is even more complicated, and mortality from *Cryptosporidium* infection may be seen in children who are undernourished or who have little access to medical care (Macfarlane and Horner-Bryce, 1987; Sarabia-Arce et al., 1990).

Diagnostic tests

The standard in laboratory diagnosis of *Cryptosporidium* is the microscopic examination of stained fecal smears. A modified acid fast (Ziehl-Neelsen) stain (Current and Garcia, 1991) is most commonly used in countries or areas with limited laboratory resources. This stain is the easiest and least expensive method available, but suffers from a detection limit of approximately 100,000 oocysts per gram feces (Weber et al., 1991). This limit may be substantially improved upon since stained oocysts are fluorescent under UV microscopy (Nielsen and Ward, 1999). Examination of slides by this method improves the detection 10-fold. The standard in the

U.S. and other developed countries is the ELISA or the immunofluorescence assay, both of which are based on a monoclonal antibody specific to the oocyst wall. This antibody reacts with both *Cryptosporidium* genotypes and many of the non-*parvum* species. The detection limit is approximately 10,000 oocysts per gram feces (Weber et al., 1991). Other methods have been developed that are designed to increase the sensitivity of the assay, but most have not been validated with clinical specimens or accepted for widespread use. These methods include flow cytometric determination of fluorescently-labeled oocysts (Valdez et al., 1997) and PCR amplification of *Cryptosporidium* gene sequences (Morgan et al., 1998). Flow cytometry has a detection limit of approximately 1000-4000 oocysts per ml, while PCR methods claim sensitivities in the range of 100 oocysts (Webster et al., 1996). Since PCR methodologies are being adopted for the diagnosis of other pathogens, it is likely that this assay will eventually become the standard in the sensitive and specific identification of *Cryptosporidium* in human specimens. This method has the added advantage of being adapted for epidemiological studies, which will need to identify *Cryptosporidium* genotype and species.

Treatment and control measures

No curative therapy beyond the host immune response is currently available despite testing of large numbers of drugs that have been approved for other purposes and a myriad of new compounds. In addition, much work is being done to identify parasite-specific, molecular targets for drug design. Past experience with a variety of agents has been reviewed (Blagburn and Soave, 1997). Several drugs, including spiramycin, initially reported as having an effect on the course of *Cryptosporidium* infections have not proven efficacious in a randomized, controlled trial (Blagburn and Soave, 1997). Paromomycin, a non-absorbable aminoglycoside, has yielded variable results ranging from no response (Hewitt et al., 2000) to a partial effect on symptoms and oocyst shedding (White et al., 1994). Disturbingly, patients often developed biliary disease during the course of the therapy. Further, in those patients who did respond, the symptoms and oocyst shedding typically returned to pre-treatment levels once the drug was discontinued. To address the spread of the parasite to the biliary tree, a systemic drug, azithromycin, was added to the regimen. In an open label study of AIDS patients with chronic cryptosporidiosis, those receiving the combination regimen experienced a decrease in stool frequency (3.0/day versus 6.5/day) and a 99 percent decrease in oocyst shedding (Smith et al., 1998). Newer drugs now appear to hold some promise. Nitazoxanide has been tested in a double-blind, placebo controlled trial in 66 Mexican AIDS patients (Rossignol et al., 1998). Two regimens, 500 mg bid or 1000 mg bid for 14 days, were associated with a complete resolution of diarrhea in 12/19

(63 percent) or 10/15 (67 percent), respectively. In another study of twelve hospitalized patients, nitazoxanide doses of 500 mg bid for 7 days resulted in a >95 percent decrease in oocysts in seven patients and complete resolution of diarrhea in four of these seven (Doumbo et al., 1997). Roxithromycin, a newer macrolide, has been used in two uncontrolled studies of AIDS associated cryptosporidiosis. A dosage of 300 mg bid for 4 weeks resulted in a complete resolution of symptoms in15/22 (68 percent) and 12/24 (50 percent) patients, respectively (Uip et al., 1998; Sprinz et al., 1998). Clinical improvement was seen in an additional 27-29 percent of patients. Large clinical trials of these agents will be necessary to establish their potential.

As stated earlier, an intact immune system is the best means of eliminating the parasite. Since there are points in the *Cryptosporidium* life cycle that require release of the parasite into the extracellular environment, it was thought that specific antibodies in the intestinal lumen might be able to neutralize the organism and assist in controlling its spread to neighboring enterocytes. A rich source of such antibodies is the colostrum from a hyperimmune cow. This material (HBC) was used in several cases of AIDS-related chronic cryptosporidiosis with varied results (reviewed in Crabb, 1998). Once again, however, when HBC or *Cryptosporidium*-specific monoclonal antibodies were subjected to controlled trials, the outcomes were somewhat disappointing in that oocyst levels and symptoms were typically decreased, but infection was not eradicated (Crabb, 1998). Likewise, bovine-derived antibodies given before and after experimental challenge with *C. parvum* oocysts failed to protect healthy volunteers from infection (Okhuysen et al., 1998). A modest decrease in symptoms and oocyst shedding was noted. Thus, to date these studies suggest that oral antibody therapy may have an effect on some patients, but the failure of these preparations in eliminating infection has tempered the enthusiasm for these treatments.

Recent advances and contemporary challenges

The widespread exposure to *Cryptosporidium* as evidenced by high seroprevalence and the severe outcome in immunocompromised populations have earned this parasite a place on the list of emerging diseases in the U.S. The EPA is taking an active role in defining the risk of acquiring *Cryptosporidium* via the drinking water supply and in recreational waters. Person to person transmission among children and secondary infections in the family are a continuing concern in day care centers and other institutional settings. And, finally, the lack of adequate treatment for severe acute cases or chronic cryptosporidiosis continues to be a problem.

Although these concerns remain unsolved, much new information has been added to the medical and scientific literature. Studies with healthy adult volunteers have revealed that this parasite is extremely infectious with

ID_{50}'s in the range of 10-1000 oocysts (Okhuysen et al., 1999; Okhuysen et al, in press) and also shows differences in the ability to cause a diarrheal illness. Prior exposure to the parasite confers at least some protection for subsequent challenge, especially to low dose exposures as would be likely under most natural conditions (Okhuysen et al, 1998; Chappell et al, 1999). Continuing volunteer studies are examining infectivity and virulence of Type 1 versus Type 2 oocysts and with *Cryptosporidium* species once thought to only infect other vertebrates. The answers to such questions will be useful in a variety of efforts to control this parasite.

The recent advances in examining the genetic polymorphism of *Cryptosporidium* have led to the recognition of two major genotypes of *C. parvum*, one of which is likely to be named a separate species. These molecular studies have also shown that at least some *Cryptosporidium* species of animals (i.e. *C. meleagridis, C. felis, C. muris,* and *C. canis*) can and do infect a small proportion of individuals, including those with intact immune systems. These studies are relatively recent and promise to add greatly to our understanding of the basic biology and pathogenicity of these interesting parasites. A *Cryptosporidium* genome project is also in progress and is in an exponential phase of increasing the genetic database (see Links). These efforts are already providing a rich source of molecular information that can be tapped for a myriad of uses.

The search for new and effective drugs continues at a fast pace and includes screening of large numbers of natural and synthetic compounds. In parallel, other studies are seeking to understand why drugs that are effective in other coccidial infections fail to have the desired effect against *Cryptosporidium.* Examination of various molecular targets is underway as are experiments designed to understand the transport of molecules into and out of the protective parasitophorous vacuole. Some new compounds are effective in *in vitro* and animal models and hold promise for future clinical studies.

Previous attempts to employ immunotherapeutic strategies have so far proven only partially successful. However, new information regarding cytokines and chemokines that are regulators and effectors of an effective immune response to *Cryptosporidium* may soon reveal ways in which the mucosal response in immunocompromised persons can be enhanced. These treatment strategies may well be a necessary adjunct to chemotherapy in severely compromised patients. A number of investigators have described molecular interactions that are necessary in the attachment and invasion process. Many of these molecules on the parasite surface can be neutralized by specific antibodies. These observations along with the recognition of an acquired protective response raise the possibility of developing an effective vaccine, which would presumably be of value in immunizing sensitive populations and in protecting or ameliorating disease of children in developing countries.

Although a vast amount of information has been gained since *Cryptosporidium* was first recognized as a human infection in 1976, we are just beginning to understand the important aspects of infection, illness, and recovery from this ubiquitous parasite. The next few years promise to hold as much or more advance in our understanding as the last 25 years.

REFERENCES

Anonymous. 1982. Cryptosporidiosis: assessment of chemotherapy of males with acquired immune deficiency syndrome (AIDS). *MMWR* 31: 589-592.
Anonymous. 1998. Foodborne outbreak of cryptosporidiosis—Spokane, Washington, 1997. *MMWR* 47: 565-567.
Bannister P, Mountford RA. 1989. *Cryptosporidium* in the elderly: a cause of life-threatening diarrhea. *Am J Med* 86: 507-508.
Besser-Wiek JW, Forfang J, Hedberg CW, et al. 1996. Foodborne outbreak of diarrheal illness associated with Cryptosporidium parvum—Minnesota, 1995. *MMWR* 45: 783-785.
Blagburn BL, Soave R. 1997. Prophylaxis and chemotherapy: Human and animal. In: Fayer R (Ed.), *Cryptosporidium and Cryptosporidiosis*. Boca Raton, FL: CRC Press.
Brandonisio O, Maggi P, Panaro MA, et al. 1993. Prevalence of cryptosporidiosis in HIV-infected patients with diarrhoeal illness. *Eur J Epidemiol* 9: 190-194.
Carr A, Marriott D, Field A, et al. 1998. Treatment of HIV-1-associated microsporidiosis and cryptosporidiosis with combination antiretroviral therapy. *Lancet* 351: 256-261.
Casemore DP, Wright SE, Coop RL. 1997. Cryptosporidiosis—human and animal epidemiology. In Fayer R (Ed.), *Cryptosporidium and Cryptosporidiosis*. Boca Raton, FL: CRC Press.
Chappell CL, Okhuysen PC, Sterling CR, et al. 1999. Infectivity of *Cryptosporidium parvum* in healthy adults with pre-existing anti-*C. parvum* serum immunoglobulin G. *Am J Trop Med Hyg* 60: 157-164.
Chieffi PP, Sens YA, Paschoalotti MA, et al. 1998. Infection by *Cryptosporidium parvum* in renal patients submitted to renal transplant or hemodialysis. *Rev Soc Bras Med Trop* 31: 333-337
Clavel A, Arnal AC, Sanchez EC, et al. 1996. Respiratory cryptosporidiosis: case series and review of the literature. *Infection* 24: 341-346.
Cordell RL, Addiss DG. 1994. Cryptosporidiosis in child care settings: a review of the literature and recommendations for prevention and control. *Pediatr Infect Dis J* 13: 310-317.
Crabb JH. 1998. Antibody-based immunotherapy of cryptosporidiosis. *Adv Parasitol* 40: 121-149.
Current WL, Garcia LS. 1991. Cryptosporidiosis. *Clin Microbiol Rev* 4: 325-328.
D'Antonio RG, Winn RE, Taylor JP, et al. 1985. A waterborne outbreak of cryptosporidiosis in normal hosts. *Ann Intern Med* 103: 886-888.
Dietz V, Vugia D, Nelson R, et al. 2000. The Foodnet Working Group. Active, multisite, laboratory-based surveillance for *Cryptosporidium parvum*. *Am J Trop Med Hyg* 62: 368-372.
Dunand VA, Hammer SM, Rossi R, et al. 1997. Parasitic sinusitis and otitis in patients infected with human immunodeficiency virus: report of five cases and review. *Clin Infect Dis* 25: 267-272.
Duombo O, Rossignol JF, Pichard E, et al. 1997. Nitazoxanide in the treatment of cryptosporidial diarrhea and other intestinal parasitic infections associated with acquired immunodeficiency syndrome in tropical Africa. *Am J Trop Med Hyg* 56: 637-639.
DuPont HL, Chappell CL, Sterling CR, et al. 1995. The infectivity of *Cryptosporidium parvum* in healthy volunteers. *N Engl J Med* 30: 855-859.

Gatei W, Ashford RW, Beeching NJ, et al. 2002. *Cryptosporidium muris* infection in an HIV-infected adult, Kenya. *Em Infect Dis* 8: 204-206.
Gentile, G, Venditti M, Micozzi A, et al. 1991. Cryptosporidiosis in patients with hematologic malignancies. *Rev Infect Dis* 13: 842-846.
Gomez-Morales MA, Ausiello CM, Guarino A, et al. 1996. Severe, protracted intestinal cryptosporidiosis associated with interferon gamma deficiency: pediatric case report. *Clin Infect Dis*23: 1336-1337.
Guarino A, Castaldo A, Russo S, et al. 1997. Enteric cryptosporidiosis in paediatric HIV infection. *J Pediatr Gastroenterol Nutr* 25: 182-187.
Hayes EB, Matte TD, O'Brien TR, et al. 1989. Large community outbreak of cryptosporidiosis due to contamination of a filtered public water supply. *N Eng J Med* 320: 1372-1376.
Hellard ME, Sinclair MI, Forbes AB, Fairley CK. 2001. A randomized, blinded, controlled trial investigating the gastrointestinal health effects of drinking water quality. *Environ Health Perspec* 109: 773-778.
Hewitt, RG, Yiannoutsos CT, Higgs ES, et al. 2000. Paromomycin: no more effective than placebo for treatment of cryptosporidiosis in patients with advanced human immunodeficiency virus infection. AIDS Clinical Trial Group. *Clin Infect Dis* 31: 1084-1092.
Jacyna MR, Parkin J, Goldin R, Baron JH. 1986. Protracted enteric cryptosporidial infection in selective immunoglobulin A and *Saccharomyces* opsonin deficiencies. *Gut* 31: 714-716.
Kandil HM, Berschneider HM, Argenzio RA. 1990. Tumour necrosis factor alpha changes porcine intestinal ion transport through a paracrine mechanism involving prostaglandins. *Gut* 35: 934-940.
Katsumata T, Hosea D, Ranuh IG, et al. 2000. Short report: Possible *Cryptosporidium muris* infection in humans. *Am J Trop Med Hyg* 62: 70-72.
Kocoshis, S.A., Cibull, M.L., Davis, T.E., Hinton, J.T., Seip, M., Banwell, J.G. Intestinal and pulmonary cryptosporidiosis in an infant with severe combined immune deficiency. J Pediatr Gastroenterol Nutr 1984; 3:149-157.
Laughon BE, Druckman DA, Vernon A, et al. 1988. Prevalence of enteric pathogens in homosexual men with and without acquired immunodeficiency syndrome. *Gastroenterology* 94: 984-993.
Leach CT, Koo FC, Kuhls TL, et al. 2000. Prevalence of *Cryptosporidium parvum* infection in children along the Texas-Mexico border and associated risks. *Am J Trop Med Hyg* 62: 656-661.
LeChevallier MW, Norton WD, Lee RG. 1991. Occurrence of *Giardia* and *Cryptosporidium* spp. In surface water supplies. *Appl Environ Microbiol* 57: 2610-2616.
Macfarlane, DE, Horner-Bryce J. 1987. Cryptosporidiosis in well-nourished and malnourished children. *Acta Paediatr Scand* 76: 474-477.
MacKenzie WR, Hoxie MJ, Proctor ME, et al. 1994. A massive outbreak in Milwaukee of *Cryptosporidium* infection transmitted through the public water supply. *N Eng J Med* 331: 161-167.
McLauchlin J, Amar C, Pedraza-Diaz S, Nichols GL. 2000. Molecular epidemiological analysis of *Cryptosporidium* spp. In the United Kingdom: results of genotyping *Cryptosporidium* spp. In 1,705 fecal samples from humans and 105 fecal samples from livestock animals. *J Clin Micro*38: 3984-3990.
Meisel JL, Perera DR, Meligro C, Rubin CE. 1976. Overwhelming watery diarrhea associated with a *Cryptosporidium* in an immunnosuppressed patient. *Gastroenterology* 70: 1156-1160.
Millard PS, Gensheimer KF, Addiss DG, et al. 1994. An outbreak of cryptosporidiosis from fresh-pressed apple cider. *JAMA* 272: 1592-1596.
Morgan UM, Pallant L, Dwyer BW, et al. 1998. Comparison of PCR and microscopy for detection of *Cryptosporidium parvum* in human fecal specimens: clinical trial. *J Clin Microbiol* 36: 995-998.

Morgan U, Weber R, Xiao L, et al. 2000. Molecular characterization of *Cryptosporidium* isolates obtained from human immunodeficiency virus-infected individuals living in Switzerland, Kenya, and the United States. *J Clin Microbiol* 38: 1180-1183.
Navin TR, Hardy AM. 1987. Cryptosporidiosis in patients with AIDS. *J Infect Dis* 155: 150.
Nielsen CK, Ward LA. 1999. Enhanced detection of *Cryptosporidium parvum* in the Acid-Fast Stain. *J Vet Diagn Invest* 11: 567-569.
Nime FA, Burek JD, Page DL, et al. 1976. Acute enterocolitis in a human being infected with the protozoan *Cryptosporidium. Gastroenterology* 70: 592-598.
Okhuysen PC, Chappell CL. *Cryptosporidium* virulence determinants. Are we there yet? *Int J Parasitol* (Suppl.) (in press).
Okhuysen PC, Chappell CL, Crabb JH, et al. 1998. Prophylactic effect of bovine anti-*Cryptosporidium* hyperimmune colostrums immunoglobulin in healthy volunteers challenged with *Cryptosporidium parvum. Clin Infect Dis*26: 1324-1329.
Okhuysen PC, Chappell CL, Crabb JH, et al. 1999. Virulence of three distinct *Cryptosporidium parvum* isolates for healthy adults. *J Infect Dis* 180: 1275-1281.
Okhuysen PC, Chappell CL, Sterling CR, et al. 1998. Susceptibility and serologic response of healthy adults to reinfection with *Cryptosporidium parvum. Infect Immun* 66: 441-443.
Okhuysen PC, Rich SM, Chappell CL, et al. Infectivity of a *Cryptosporidium parvum* isolate of cervine origin for healthy adults and gamma interferon knock out mice. *J Infect Dis* (in press).
Patel S, Pedraza-Diaz S, McLauchlin J, Casemore DP. 1997. Molecular characterization of *Cryptosporidium parvum* from two large suspected waterborne outbreaks. Outbreak control Team South and West Devon 1995, Incident Management Team and Further Epidemiological and Microbiological Studies Subgroup North Thames 1997. *Com Dis Publ Hlth* 1: 231-233.
Pedraza-Diaz S, Amar C, Iversen AM, et al. 2001. Unusual *Cryptosporidium* species recovered from human faeces: first description of *Cryptosporidium felis* and *Cryptosporidium* "dog type" from patients in England. *J Med Microbiol* 50: 293-296.
Pedraza-Diaz S, Amar C, McLaughlin J. 2000. The identification and characterization of an unusual genotype of *Cryptosporidium* from human faeces as *Cryptosporidium meleagridis. FEMS Microbiol Lett* 189: 189-194.
Pieniazek, NJ, Bornay-Llinares FJ, Slemenda SB, da Silva AJ, et al. 1999. New *Cryptosporidium* genotypes in HIV-infected persons. *Emerg Inf Dis* 5: 444-449.
Quiroz ES, Bern C, MacArthur JR, et al. 2000. An outbreak of cryptosporidiosis linked to a foodhandler. *J Infect Dis* 181: 695-700.
Rene, E, Marche C, Regnier B, et al. 1989. Intestinal infections in patients with acquired immunodeficiency syndrome. A prospective study in 132 patients. *Dig Dis Sci* 34: 773-780.
Rivasi F, Rossi P, Righi E, Pozio, E. 1999. Gastric cryptosporidiosis: correlation between intensity of infection and histological alterations. *Histopathology* 34: 405-409.
Rochelle PA, De Leon R. 2001. A review of methods for assessing the infectivity of *Cryptosporidium parvum* using in-vitro cell culture. In: Smith M, Thompson KC (Eds.), *Cryptosporidium: The Analytical Challenge.* Royal Society of Chemistry.
Rossignol JF, Hidalgo H, Feregrino M, et al. 1998. A double-'blind' placebo controlled study of nitazoxanide in the treatment of cryptosporidial diarrhoea in AIDS patients in Mexico. *Trans Roy Soc Trop Med Hyg* 92: 663-666.
Sarabia-Arce S, Salazar-Lindo E, Gilman RH, et al. 1990. Case-control study of *Cryptosporidium parvum* infection in Peruvian children hospitalized for diarrhea: possible association with malnutrition and nosocomial infection. *Pediatr Infect Dis J* 9: 627-631.
Smith NH, Cron S, Valdez LM, et al. 1998. Combination drug therapy for cryptosporidiosis in AIDS. *J Infect Dis* 178: 900-903.
Smith PD, Lane HC, Gill VJ, et al. 1988. Intestinal infection in patients with the acquired immunodeficiency syndrome (AIDS): etiology and response to therapy. *Ann Intern Med* 108: 328-333.

Soave R, Armstrong D. 1986. *Cryptosporidium* and cryptosporidiosis. *Rev Infect Dis* 8: 1012-1023.
Sorville FJ. 1990. Swimming-associated cryptosporidiosis—Los Angeles County. *MMWR* 39: 343-345.
Sprinz E, Mallman R, Barcellos S, et al. 1998. AIDS-related cryptosporidial diarrhoea: an open study with roxithromycin. *J Antimicrob Chemother* 41 Suppl B.: 85-91.
Tankyüksel M, Gün H, Doganci L. 1995. Prevalence of *Cryptosporidium* sp. In patients with neoplasia and diarrhea. *Scand J Infect Dis* 27: 69-70.
Thielman NM, Guerrant RL. 1998. Persistent diarrhea in the retuned traveler. *Infect Dis Clin North Am* 12: 489-501.
Tyzzer EE. 1912. *Cryptosporidium parvum* (sp. nov.), a coccidium found in the small intestine in the common mouse. *Arch Protistenkd* 26: 394-418.
Uip DE, Lima AL, Amato VS, et al. 1998. Roxithromycin treatment for diarrhoea caused by *Cryptosporidium* spp. In patients with AIDS. *J Antimicro Chemother* 41Suppl. B: 93-97.
Ungar BLP. 1990. Cryptosporidosis in humans (*Homo sapiens*). In: Dubey JP, Speer CA, Fayer R (Eds.), *Cryptosporidiosis of Man and Animals.* Boca Raton, FL, CRC Press. ..
Valdez LM, Dang H, Okhuysen PC, Chappell CL. 1997. Flow cytometric detection of *Cryptosporidium* oocysts in human stool samples. *J Clin Microbiol* 35: 2013-2017.
Weber R, Bryan RT, Bishop HS, et al. 1991. Threshold detection of *Cryptosporidium* oocysts in human stool specimens: evidence for low sensitivity of current diagnostic methods. *J Clin Microbiol* 29: 1323-1327.
Webster KA, Smith HV, Giles M, et al. 1996. Detection of *Cryptosporidium parvum* oocysts in faeces: comparison of conventional coproscopical methods and the polymerase chain reaction. *Vet Parasitol* 61: 5-13.
White AC, Jr, Chappell CL, Hayat CS, et al. 1994. Paromomycin for cryptosporidiosis in AIDS: A prospective, double-blind trial. *J Infect Dis* 170: 419-424.
Widmer G, Akiyoshi D, Buckholt MA, et al. 2000. Animal propagation and genomic survey of a genotype 1 isolate of *Cryptosporidium parvum*. *Mol Biochem Parasitol* 108: 187-197.
Widmer G, Tzipori S, Fichtenbaum CJ, Griffiths JK. 1998. Genotypic and phenotypic characterization of *Cryptosporidium parvum* isolates from people with AIDS. *J Infect Dis* 178: 834-840.
Xiao L, Bern C, Limor J, et al. 2001. Identification of 5 types of *Cryptosporidium* parasites in children in Lima, Peru. *J Inf Dis* 183: 492-497.

Links to *Cryptosporidium* Genome Projects

Cambridge project, www.mrc-lmb.cam.ac.uk/happy/CRYPTO/crypto-genome.html
Genotype 1 sequencing project, www.parvum.mic.vcu.edu
Genotype 2 sequencing project, www.cbc.umn.edu/Research Projects/AGAC/Cp/index.html
EST project, www.medsfgh.ucsf.edu/id/CpTags/home.html

TOXOPLASMOSIS

David S. Lindsay[1], Louis M. Weiss[2], and Yasuhiro Suzuki[1]

[1]Center for Molecular Medicine and Infectious Diseases, Department of Biomedical Sciences and Pathobiology, Virginia-Maryland Regional College of Veterinary Medicine, Virginia Tech, Blacksburg, Virginia
[2]Departments of Medicine and Pathology, Albert Einstein College of Medicine, Bronx, New York

Etiologic Agent: *Toxoplasma gondii* (Nicolle and Manceaux, 1908). *Toxoplasma gondii* is a significant foodborne pathogen and the third leading cause of death from foodborne disease in the United States (Mead et al., 1999).
Clinical Manifestations: Clinical symptoms vary depending on the age and immune status of the patient. About 15 percent of cases in the United States are symptomatic (Mead et al., 1999). Infections in congenitally infected patients and others with impaired immune functions are usually much more severe than in immunocompetent patients. Acute disease and death can occur in a primary infection. Re-activated toxoplasmosis occurs in individuals that have a latent infection and subsequently become immunosuppressed.
Complications: Depend on the age and immune status of the patient.
Mode Of Transmission: Humans become infected by ingesting tissue cysts containing bradyzoites in raw or undercooked fresh meat or poultry products or by ingesting sporulated oocysts in contaminated water, or contaminated raw produce or by ingesting contaminated soil. A seronegative pregnant female can transmit the infection to her fetus if she acquires infection during pregnancy. Infections can also be acquired by tissue transplantation.
Diagnosis: Diagnosis is made by identification of typical clinical signs and serology in most cases. PCR can be used as a diagnostic tool on clinical samples. Immunohistochemial tests can be done on biopsies and autopsy specimens.
Drug(s) Of Choice: Pyrimethamine (50 to 100 mg daily) plus sulfadiazine (4 to 8 g daily) is the drug combination of choice for acute infections and recrudescent disease. Since no drugs will destroy the latent tissue cyst stage, recrudescence is a problem.
Reservoir Hosts: Cats and other felids can excrete oocysts in their feces. Tissue cysts can be found in most mammalian and avian species.
Control Measures: Prevention of exposure to viable tissue cysts in meat products and prevention of exposure to oocysts on produce, in the soil or in drinking water.

ETIOLOGY/NATURAL HISTORY

Toxoplasma gondii (Nicolle and Manceaux, 1908) is in the protozoan Phylum Apicomplexa. This group of obligatory parasites contains the malarial organisms and other important pathogens of humans and domestic animals. Toxoplasmosis was originally described in 1908 from the gondi, (*Ctenodactylus gundi*), a North African rodent used in research at the Pasture Institute in Tunis. The gondi is therefore the type intermediate host. The complete life cycle of *T. gondii* was not fully described until 1970 when domestic cats were shown to be definitive hosts and excrete oocysts in their feces. The first case of human toxoplasmosis was reported in 1923 in an 11-month-old congenitally infected infant that had hydrocephalus and microphthalmia with coloboma (Remington et al., 1995). In the late 1930's and early 1940's it became well established that toxoplasmosis is an important disease of humans and that infections in infants were acquired prenatally. The percentage of congenital toxoplasmosis cases in humans was too low to explain the high seroprevalence of *T. gondii* in the populations examined. Carnivorism was suggested by several researchers and conclusively proven in 1965. Ingestion of infected meat, however, did not explain *T. gondii* infection in vegetarians or herbivores and other modes of transmission had to be present. Hutchison first found resistant *T. gondii* in cat feces in 1965 and thought it was enclosed in the eggs of *Toxocara cati* the feline ascarid (Dubey and Beattie, 1988). Several studies disproved the association of *T. gondii* with *Toxocara cati* eggs. In 1969-1970 several groups of researchers reported the presence of a small coccidial oocyst in cat feces that was *T. gondii*. *Toxoplasma gondii* oocyst excretion has since been observed in several species of felids including bobcats, mountain lions, ocelots, Bengal tigers, margays, jaguarondis, and Iriomote cats.

The life cycle of *T. gondii* is complex (Figure 1). Stages found in both the intermediate and definitive hosts are bradyzoites in tissue cysts (Figure 2) and tachyzoites (Figure 3) in tissues. Bradyzoites (brady = slow) are slowly dividing stages responsible for latent infections, and tachyzoites (tachy = fast) are rapidly dividing stages responsible for disseminating the infection. Schizonts, merozoites, and sexual stages all are present in enterocytes of the definitive feline host. The life cycle is best explained by describing events that occur in intermediate hosts and those that occur in the primary host.

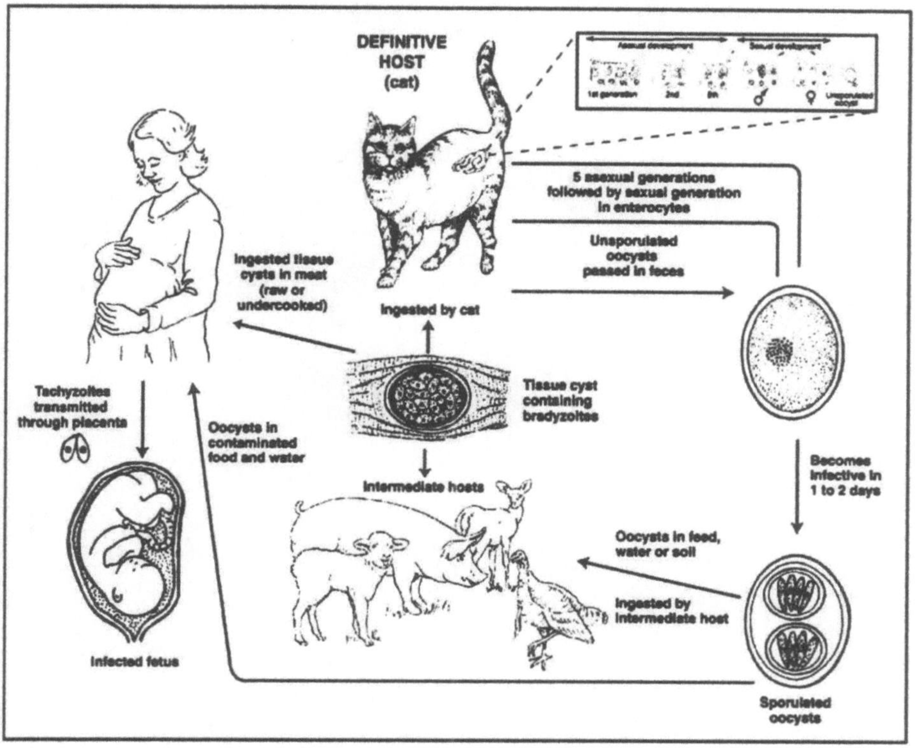

Figure 1. Life cycle of Toxoplasma gondii.

Life Cycle in the Intermediate Host

When meat containing tissue cysts is ingested, the bradyzoites are liberated from tissue cysts after exposure to the acid conditions of stomach. Bradyzoites penetrate the mucosa of the small intestine and begin multiplication in cells in the lamina propria. The bradyzoites convert to tachyzoites within a few days (Bohne et al., 1993; Soete et al., 1993) and the tachyzoites are disseminated throughout the body by the lymphatic and vascular systems. Tissue cysts containing bradyzoites are then produced in a variety of tissues including the heart, retina, and skeletal muscles. These tissue cysts remain viable for several years to the life of the individual.

When sporulated oocysts are ingested, sporozoites are liberated from the sporocysts within the oocysts into the lumen of the duodenum. Sporozoites penetrate the mucosa of the small intestine and begin multiplication in cells in the lamina propria as tachyzoites. The tachyzoites are disseminated throughout the body by the lymphatic and vascular systems. Tissue cysts containing bradyzoites are then produced in a variety of tissues.

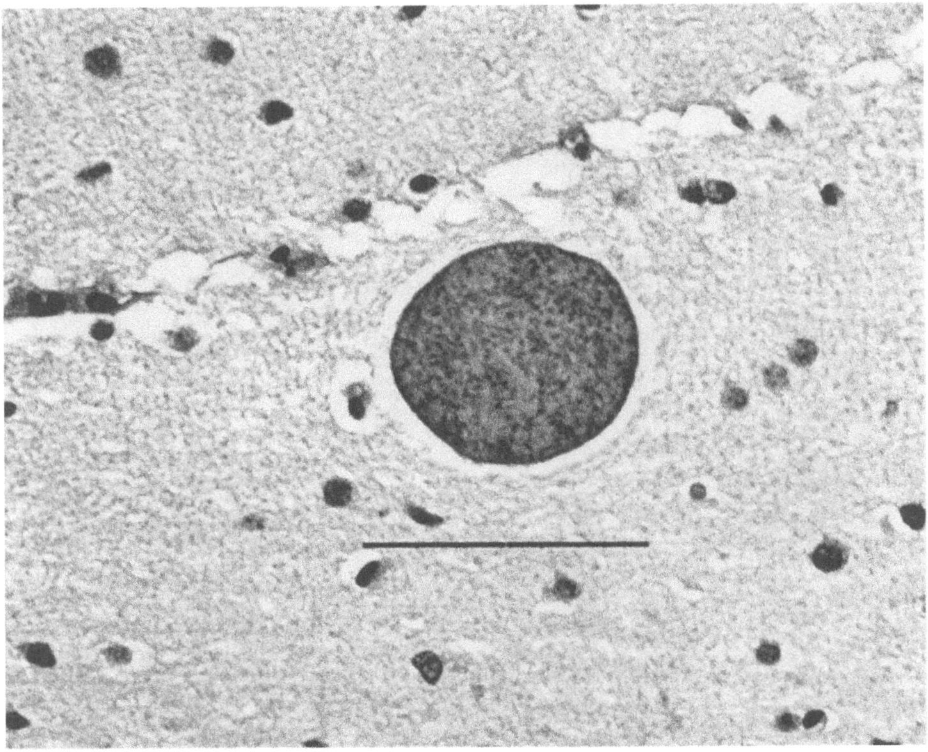

Figure 2. Tissue cyst of Toxoplasma gondii *in the brain of a mouse. Hematoxylin and Eosin stain. Bar = 50 micrometers*

Life Cycle in the Cat

In the cat *T. gondii* undergoes a coccidial like cycle in enterocytes of the small intestine. Five distinct asexual schizont types are present in enterocytes of the feline intestine and they produce the sexual stages that give rise to the oocysts (Dubey and Frenkel, 1972). Oocysts are excreted in the feces for several weeks but high numbers are only excreted during the first week of patency (Figure 4). Oocysts then develop (sporulate) outside in the environment in 2 to several days depending on environmental conditions. The sporulated infective oocyst has 2 sporocysts that each enclose 4 infective sporozoites.

The length of time it takes before oocysts are excreted (prepatent period) varies depending on the stage of *T. gondii* that is ingested by the cat (Freyre et al, 1989). Oral inoculation of bradyzoites/tissue cysts is most efficient in inducing oocyst-producing infections in cats with 97 percent excreting oocysts with a short 3-6 day prepatent period (Dubey and Frenkel, 1976).

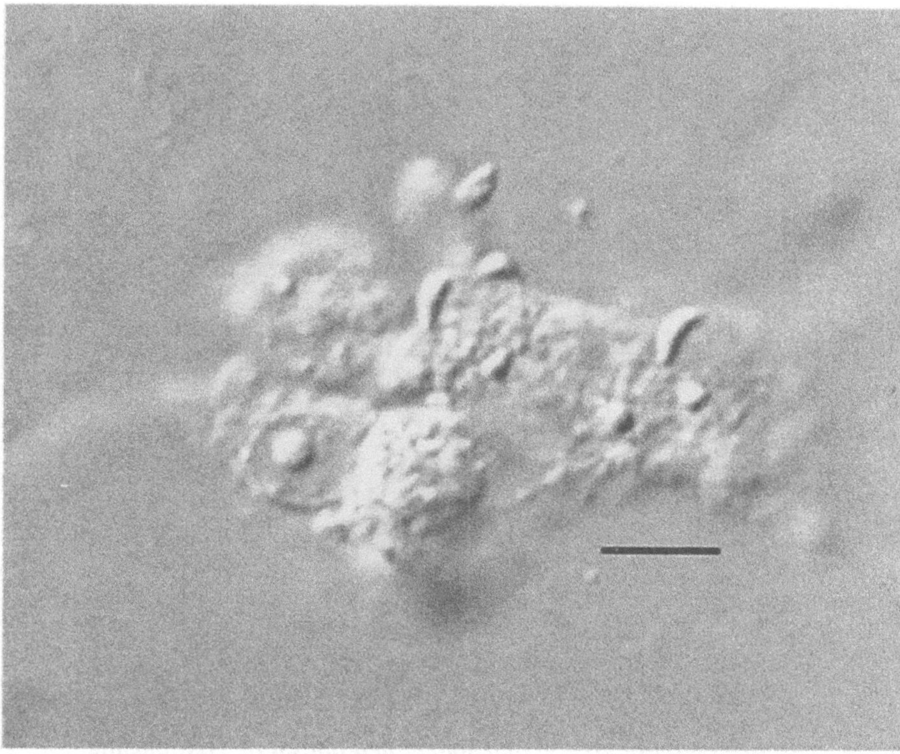

Figure 3. Tachyzoites of Toxoplasma gondii *(arc shaped bodies) in mouse macrophages. Unstained. Bar = 10 micrometers*

Oral inoculation of tachyzoites or oocysts is less efficient because only the bradyzoite stage can undergo development in the enterocytes and produce the stages that eventually terminate in oocyst excretion. Oral inoculation of cats with oocysts or tachyzoites produces oocyst-excreting infections in only 16 percent and 20 percent of cats, respectively, and the prepatent period of these infections is 21-40 days (Dubey and Frenkel, 1976; Freyre et al., 1989).

Cats and *T. gondii* Oocyst Biology

Cats are pivotal in the transmission of the parasite to herbivores and omnivores. It is important to emphasize that only the bradyzoite stage from the tissue cyst can cause the enteroepithelial cycle in the cat's enterocytes, which terminates in oocysts excretion in the feces. All ages, sexes, and breeds of domestic cats are susceptible to *T. gondii* infection. Transplacentally or lactogenically infected kittens will excrete oocysts but the prepatent period is usually 3 weeks or more. Kittens are infected with tachyzoites that necessitates a stage conversion to bradyzoite and migration to the intestine. Domestic cats under 1 year of age produce the most numbers

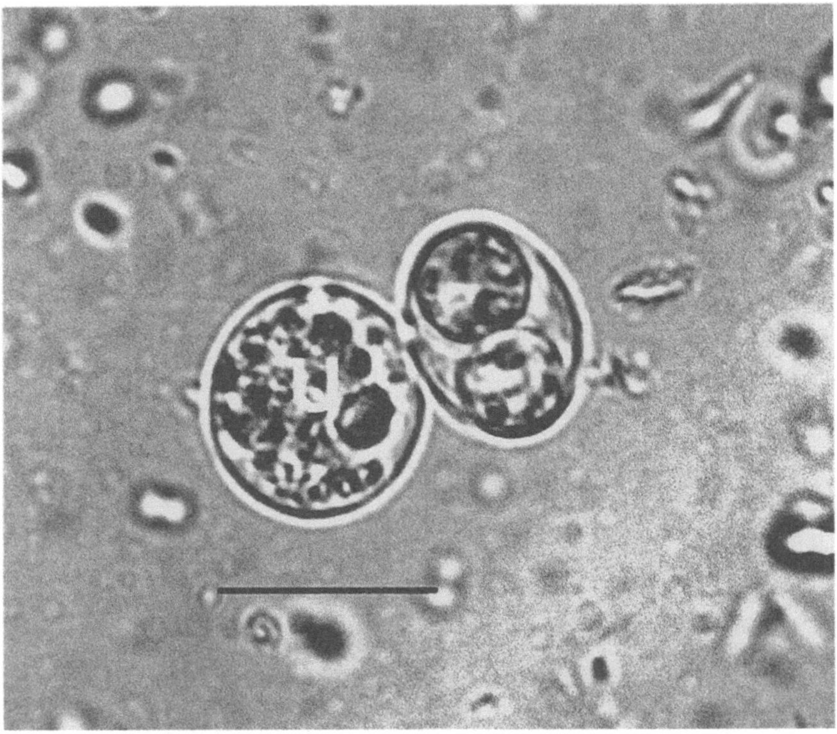

Figure 4. Unsporulated oocyst (U) and sporulated oocyst of Toxoplasma gondii *in cat feces. Unstained. Bar = 10 micrometers*

of *T. gondii* oocysts. Cats that are born and raised outdoors usually become infected with *T. gondii* shortly after they are weaned and begin to hunt. *Toxoplasma gondii* naive adult domestic cats will excrete oocysts if fed tissue cysts but they usually will excrete fewer numbers of oocysts and excrete oocysts for a shorter period of time than recently weaned kittens.

Unsporulated *T. gondii* oocysts are spherical to subspherical, and contain a single mass (sporont). Sporulation occurs in the environment and is dependent on temperature and moisture (Dubey et al., 1970a). Sporulation is asynchronous and some oocysts will sporulate before others. Completely infectious oocysts are present by 24 hr at 25 C (room temperature); by 5 days at 15 C, and by 21 days at 11 C (Dubey et al., 1970b). Unsporulated oocysts do not survive freezing but can remain viable at 4 C for several months and become infectious if placed under the appropriate conditions. Unsporulated oocysts die if kept at 37 C for 24 hours and are killed by 10-minute exposure to 50 C. They will survive for up to 6 weeks if kept at 4 C (Lindsay et al., 2002). A small population of unsporulated oocysts can survive anaerobic conditions for 30 days and remain infectious. Oocysts do not sporulate in 0.3 percent formalin, 1 percent ammonium hydroxide solution or in 1 percent iodine in 20 percent ethanol but can sporulate in 5 percent sulfuric acid, 20 percent ethanol, 10 percent ethanol plus 10 percent ether, 1 percent hydrochloric acid, 1 percent phenol and in tap water (Dubey et al., 1970a, 1970b). Drying kills *T. gondii* oocysts.

Sporulated *T. gondii* oocysts are subspherical to ellipsoidal and each contains 2 ellipsoidal sporocysts, which enclose 4 sporozoites. Sporulated oocysts are more resistant to environmental and chemical stresses than are unsporulated oocysts. Viable oocysts of *T. gondii* have been isolated from soil samples (Ruiz et al., 1973; Coutinho et al., 1982; Frenkel et al., 1995) and experimentally can survive for over 18 months in the soil (Frenkel et al., 1975). Sporulated oocysts cannot survive freezing or temperatures of 55 C or greater (Dubey, 1998). Sporulated oocysts survive for several years at 4 C in liquid medium (Dubey, 1998). Sporulated oocysts can survive for at least 8 weeks on the surface of raspberries and blueberries kept at 4 C (Kneil et al., 2002).

Irradiation with >0.4 to 0.8 kGy of gamma radiation inactivates nonsporulated oocysts of *T. gondii* (Dubey et al., 1996). Irradiation with 0.4 kGy of gamma radiation reduces the infectivity of sporulated oocysts of *T. gondii* when placed on raspberries and fed to mice but does not completely prevent infection (Dubey et al., 1998).

Cockroaches, flies, earthworms and other phoretic hosts can serve to distribute *T. gondii* oocysts from the site of feline defecation in the soil.

Miscellaneous Aspects of *T. gondii* Biology

The tachyzoite and tissue cyst stages of *T. gondii* are safe and easy to work with in the laboratory. These stages cannot osmoregulate and cannot survive for long outside of host cells. Simple washing with water will kill these stages. The oocyst stage is the most difficult to work with in the laboratory for several reasons. Cats must be used as definitive hosts to obtain the oocysts, the oocysts must be separated form cat feces and sporulated, the oocysts are long-lived and resistant to disinfectants, and accidental contamination of researchers and non-target laboratory animals is always a possibility. Researchers probably become infected with oocysts during the process of purifying sporulated oocysts from cat feces or when conducting experimental studies using oocysts. Sporulated oocysts of *T. gondii* can pass through the intestinal tract of animals and remain infective (Lindsay et al., 1997). These oocysts can contaminate animal bedding and can pose a threat to animal caretakers and potentially to other animals. Consequently, it is best to collect animal bedding for the first week after infection and autoclave it to ensure that any contaminating oocysts are killed.

Continuous passage of *T. gondii* in mice or in cell cultures can alter its developmental biology. The ability to produce oocyst-excreting infections in the feline intestine is lost after 30-35 serial passages in mice (Frenkel et al.,

1976) or 40 in cell culture (Lindsay et al., 1991). Highly virulent strains of *T. gondii*, such as the RH strain (Sabin, 1941), have been referred to as being "tissue cyst-less" but this is a misconception because experimental mouse hosts die before tissue cysts can be produced. The nonpathogenic and non-persistent temperature sensitive mutant Ts- 4 strain (Pfefferkoen and Pfefferkoen, 1976) of *T. gondii* is the only truly tissue cyst- less strain of *T. gondii*.

Most strains of *T. gondii* will produce tissue cysts in cell culture with no external manipulation (Lindsay et al., 1993). Sporozoites, tachyzoites or bradyzoites can all produce tissue cysts in cell cultures. The tissue cysts that are produced in cell culture are biologically similar to those produce in host tissue and can produce oocysts excreting infections in cats (Lindsay et al., 1991). Cell culture systems can be manipulated by a variety of chemical and thermal methods to drive the tachyzoite to bradyzoites stage conversion process and produce an increased production of tissue cysts in cell culture (Weiss et al., 1995).

EPIDEMIOLOGY

Toxoplasma gondii has a worldwide distribution. Infections are widespread in mammals and avian species. Studies conducted in the United States on serum collected from 17,658 individuals over 12 years of age found that 22.5 percent were positive for IgG antibodies. The prevalence of *T. gondii* antibody among women of childbearing age was 15 percent (Jones et al., 2001). These findings indicate that the United States has a low prevalence rate and that 85 percent of women of childbearing age are at risk of developing congenital toxoplasmosis.

Domestic cats are important in the transmission of *T. gondii* and its maintenance in the environment. As noted previously, young domestic cats excrete the largest numbers of *T. gondii* oocysts in their feces. There are more than 50 million cats in the United States (Patronek, 1998). The prevalence of antibodies to *T. gondii* is about 58 percent in free-roaming cats and 37 percent in pet cats (Dubey, 1994). It is logical to assume that veterinarians, who have more exposure to cats (both sick and healthy) than the general public, would have a higher seroprevalence of antibodies to *T. gondii*. Serologic studies have shown that veterinarians do not have a higher seroprevalence than do non-veterinarians, however (Behymer et al., 1973; Sengbusch and Sengbusch, 1976; DiGiacomo et al., 1990). Because AIDS patients are highly susceptible to infection, a study was done in this population to determine if cat ownership had any influence on the development of toxoplasmosis. It was conclusively shown that owning cats did not increase the risk of AIDS patients developing toxoplasmosis (Wallace et al., 1993).

Humans can become infected with oocysts when they ingest food, water, or other consumable products that have been contaminated with oocysts (Sulzer et al., 1986; Isaac-Renton et al., 1998; Aramini et al., 1999). Ingestion of unwashed raw fruits or vegetables is associated with an increased risk of maternally acquired toxoplasmosis (Kapperud et al., 1996). Contact with soil is also associated with an increased risk of maternal toxoplasmosis (Decavalas et al., 1990). *Toxoplasma gondii* oocysts have been isolated from soil obtained from gardens (Coutinho et al., 1982).

Pork is the most likely commercial food source of tissue cysts for people in the United States because cattle are naturally resistant and other *T. gondii* infected meats such as sheep and goat are not consumed in significant amounts in the United States (Dubey, 1994). Chickens are susceptible to *T. gondii* infection but because chicken is often frozen and seldom eaten rare, it is not considered a primary source of infection. Wild game is also a potential source of *T. gondii*. *Toxoplasma gondii* is prevalent in commonly hunted animals, such as, white-tailed deer, wild turkeys, bear and wild pigs. Clinical toxoplasmosis has been associated with the consumption of venison (Sacks et al., 1983; Ross et al., 2001).

PATHOGENESIS

The development of clinical disease following *T. gondii* infection is dependent on many factors. The route of inoculation and parasite stage inoculated can influence the resulting *T. gondii* infection. For example, it appears that oral infections with oocysts can cause fatal enteritis in mice more often than do oral infections with tissue cysts. Tachyzoites administered intraperitoneally are usually more pathogenic that the same numbers of tachyzoites given subcutaneously. For the less virulent strains (see below), the pathogenesis of the infection increases as the dose of parasites increases. Host genetics also influence the pathogenesis of a *T. gondii* infection (McLeod et al., 1994: Luder and Seeber, 2001). Interferon-gamma dependent cell mediated immunity is needed to control *T. gondii* in acute and chronic infections (Suzuki, 2002). There are three distinct but closely related lineages of *T. gondii* that occur in nature (Howe and Sibley, 1995). Virulence correlates with parasite genotype (Sibley and Boothroyd, 1992). Type I strains are highly pathogenic for mice and usually one infectious tachyzoite will cause death. Type I strains also are frequently observed in cases of congenital and ocular toxoplasmosis indicating that they also may be likely to cause clinical disease in humans (Fuentes et al., 2001) (Grigg and Boothroyd, 2001). Type II and type III strains are less pathogenic for mice. It usually takes greater than 1,000 tachyzoites to cause fatal infections in mice with these strains. Type I strains are relatively rare in animals while Type II and III strains are more abundant. Type II strains are common in AIDS patients who develop clinical toxoplasmosis (Fuentes et al., 2001).

The differences in virulence of the three strains does not appear to be caused by different growth rates of the parasites in host cells (Dobrowolski and Sibley, 1996). The highly virulent Type I strains grow only slightly faster than the Type II or Type III strains. Differences in virulence also are not caused by differences in tissue dissemination patterns or tissue burdens of the *T. gondii* strains (Mordue et al., 2001). Recent studies indicate that acute virulence of Type 1 strains occurs by a paradoxical mechanism. Infection with virulent strains induces and overproduction of Th-1 cytokines which are normally protective but the unregulated production of these factors (primarily IL-18 but also IL-12, IFN-gamma, and TNF-alpha) leads to lethality (Sibley et al., 2002). The actual cause of death is unknown but it is speculated that increased levels of these cytokines leads to increased vascular permeability, which results in multiple organ failure and death (Sibley et al., 2002). Following oral infection of mice with lethal doses of Type II strains intestinal necrosis occurs in association with mortality. The observed lesions do not appear to be caused by tissue destruction by *T. gondii* but by Th-1-type immune responses induced by *T. gondii*. Overproduction of nitric oxide by nitric oxide synthase, stimulated by IFN-gamma and TNF-alpha, appear to cause the necrosis.

CLINICAL MANIFESTATIONS AND COMPLICATIONS

While *Toxoplasma gondii* infects a large percentage of the population the majority of these infections are asymptomatic. Most immunocompetent patients have, at most, symptoms of a viral illness during primary infection with this protozoan pathogen. The most common symptomatic presentations of *T. gondii* in an immunocompetent host are lymphadenitis or chorioretinitis. Rarely myocarditis, polymyositis or encephalitis may occur. Two groups of patients, however, are at high risk for severe disease that can result in death. These are congenitally infected newborns and patients with impaired cell mediated immunity. In these patients central nervous system infection with encephalitis or chorioretinitis is the most common manifestation.

It is believed that only 10 percent of cases of *T. gondii* infection in adults and children are symptomatic (Remington, 1974). A mononucleosis-like syndrome with associated cervical lymphadenopathy is the most common presentation, but any lymph node can be involved. Toxoplasmosis may account for 5 percent of clinically significant lymphadenopathy cases (McCabe et al., 1987). A mononucleosis-like syndrome consisting of fever, malaise, night sweats, myalgias, sore throat, hepatosplenomegaly or small numbers of circulating atypical lymphocytes may be seen. Signs and symptoms usually resolve within a few months to a year. Affected lymph nodes are firm, non-fixed to the tissue and do not suppurate. Biopsy demonstrates follicular hyperplasia, epitheloid histocytes blurring the margins of the germinal centers, focal distention of sinuses with monocytoid cells, and occasionally cysts or tachyzoites of *T. gondii.* Acquired infection occasionally may be associated with myositis or a sepsis-like syndrome.

Congenital infection results when a seronegative pregnant woman acquires *T. gondii* infection. Women who are seropositive before conception, as a rule, do not transmit infection to the fetus. There have, however, been a few rare cases reported of congenital infection where the mother acquired infection 1 to 2 months before conception (Vogel et al., 1996). In addition, immunocompromised women such as those experiencing HIV infection or treated with corticosteroids) may transmit latent infections during pregnancy resulting in congenital infection (Mitchell et al., 1990). The risk of transmission increases with each trimester, e.g. the risk of transmitting infection to the fetus is lower if maternal infection is acquired in the first trimester than in the third trimester. The severity of disease in the fetus, however, decreases with each trimester in which primary infection occurs, e.g. congenital infection in the first trimester can result in death of the fetus or severe symptoms at birth whereas infection of the fetus in the third trimester is often asymptomatic at birth (Kroppe et al., 1986; Wilson et al., 1980). Symptoms of congenital infection can include chorioretinitis, strabismus, epilepsy, mental retardation, anemia, jaundice, rash, thrombocytopenia, encephalitis, microcephaly, sensorineural hearing loss, intracranial calcifications, or hydrocephalus (Couvreur et al., 1984). Children who are asymptomatic at birth may develop chorioretinitis later in life (Wilson et al., 1980).

Toxoplasma gondii is an important cause of chorioretinitis and accounts for at least 25 percent of posterior uveitis cases in the United States (Holland et al., 1996). Ocular lesions may result from either congenital or acquired infection (Montoya and Remington, 1996; Glasner et al., 1992). The involvement of the eye is similar in both types of infection. Bilateral involvement has been seen in acquired infection. Congenital infection usually presents by age 30, whereas acquired infections are most commonly recognized in patients between the ages of 40 and 70. Chorioretinitis due to *T. gondii* is a relapsing disease that occurs following reactivation of latent infection. During reactivation bradyzoites in cysts transform to tachyzoites that proliferate and cause acute chorioretinitis that is associated with symptoms of blurred vision, scotoma, pain, photophobia and epiphora.

In individuals with impaired cell mediated immunity, latent *Toxoplasma gondii* infection can reactivate causing encephalitis. During HIV infection this most often occurs when the CD4+ count is less than 100. Infection usually presents as focal CNS lesions with associated altered mental state, seizures, weakness, neuropsychiatric manifestations, or stroke-like presentations. Spinal cord lesions also can occur. Chorioretinitis is rarely seen. Pulmonary disease due to *T. gondii* with associated cough and fever is the most common extraneural presentation of infection (Campagna, 1997). Autopsy series suggest that myocardial involvement also is common, but not recognized before death. Patients with organ transplants or malignancy also may experience central nervous system, myocarditis or pulmonary involvement due to reactivation disease (Israelski et al., 1993).

DIAGNOSTIC TESTS

The diagnosis of acute toxoplasmosis may be established by: 1) detection of anti-*T gondii* antibodies by serological tests or 2) detection of tachyzoites or *T. gondii*-specific DNA in body fluids or tissue samples. In most cases of toxoplasmosis in immunocompetent individuals, diagnosis is established by serological testing.

Detection of *T. gondii* Antibodies by Serological Tests

Demonstration of anti-*T. gondii* IgG antibody provides evidence for infection with the parasite. The presence of IgM antibody, in addition of the IgG antibody, suggests a recently acquired infection. IgM antibodies may remain detectable for more than one year after initial infection, however (Naot et al., 1982; van Loon et al., 1983). False-positive reactions in several commercial test kits also complicate the diagnosis (Liesenfeld et al., 1997; Wilson et al., 1997). Thus, for interpreting a positive IgM result, confirmatory testing should always be performed. Direct agglutination tests using acetone- and formalin-fixed tachyzoites (AC/HS test) has been shown to be effective in determining whether the infection was acquired recently or in the more distant past (Dannemann et al., 1990). A combination of the Sabin-Feldman dye test, IgM ELISA, IgA ELISA, IgE ELISA and AC/HS test can be used to differentiate between recently acquired and chronic infection (Liesenfeld et al., 1997). ELISA tests with several recombinant *T. gondii* antigens also appears to be helpful for diagnosis of recently acquired infection (Li et al., 2000).

Detection of *T. gondii* Stages or Specific Tachyzoite DNA in Samples

Isolation of *T. gondii* from blood or body fluids (e.g. CSF, amniotic, or BAL fluids) establishes diagnosis of the acute infection. The isolation can be performed by inoculation of the samples in mice or tissue culture. Demonstration of tachyzoites in histological sections or smears of body fluids by immunoperoxidase staining with anti-*T. gondii* antibodies also establishes the diagnosis (Conley et al., 1981).

Detection of *T. gondii* tachyzoite DNA in body fluids and tissues by PCR amplification is effective for diagnosing congenital (Grover et al., 1990), ocular (Montoya et al., 1999) and cerebral toxoplasmosis (Holliman et al., 1990).

TREATMENT AND CONTROL MEASURES

For purposes of therapy it is helpful to separate toxoplasmosis into several categories (Table I). The decision to treat is based on the location of infection, immune status of the patient and whether or not a woman with acute toxoplasmosis is pregnant. There are virtually no large, well-controlled clinical trials to establish optimal therapy, however, there have been several studies of prophylaxis and treatment for toxoplasmosis in the setting of HIV infection or congenital disease. The recommended therapies are based on extrapolations from *in vitro* and animal models (mostly murine) and the clinical experience and practice at medical centers experienced in the treatment of *T. gondii* infection. In general the drugs used to treat toxoplasmosis are active against the rapidly replicating tachyzoite stage, but have limited efficacy against tissue cysts. Thus, patients treated for toxoplasmosis generally will retain a latent infection (tissue cysts) at the conclusion of treatment.

TREATMENT REGIMENS

Asymptomatic Infection or Latent Infection

Immunocompetent individuals with latent toxoplasmosis, as evidenced by positive serology, do not require treatment. In AIDS patients who are seropositive for *T. gondii* the risk of developing encephalitis has been estimated at 10 to 20 percent. Trimethomprim-sulfamethoxazole is effective in preventing encephalitis in this setting (Bozzette et al., 1995), dapsone plus pyrimethamine is an alternative that is also effective (Torres et al., 1993). In cardiac transplantation, prophylaxis with pyrimethamine for 6 weeks is used for *T. gondii* seronegative recipients receiving hearts from seropositive donors.

Acquired Toxoplasmosis

Treatment is rarely needed in immunocompetent individuals. For the rare patient whose symptoms are persistent, treatment should be as described for disseminated disease. Myocarditis, encephalitis, a sepsis syndrome with shock and hepatitis are occasionally seen. In such patients treatment should consist of pyrimethamine (100 mg loading dose and 25 -50 mg per day) and sulfadiazine or trisulfapyrimidines (4 to 8 gm per day) for 4 to 6 weeks. Folinic acid (5 to 10 mg/d) also should be given. Infections acquired through a laboratory accident or blood transfusion also should be treated as described above.

Table 1: Toxoplasma infections and treatment indications

Syndrome	Treatment[a]
1. Asymptomatic Infection Latent infection detected by positive serologic testing	Not Required
2. Adenopathy, fever or malaise in the immunocompetent host	Not Required[b]
3. Disseminated disease (i.e. CNS, cardiac or liver) in the normal host or Laboratory infection with tachyzoites	PYR/SULFA
4. Infection during Pregnancy	SPR - PYR/SULFA[c]
5. Congenital Toxoplasmosis	PYR/SULFA[d]
6. Ocular Toxoplasmosis	PYR/SULFA & Steroids
7. Infection in Immunocompromised Hosts	
A) AIDS	PYR/SULFA
B) Transplantation	PYR/SULFA
C) Acute Disease	PYR/SULFA

a. Recommended primary treatments as described in the text; PYR-pyrimethamine, SULFA- sulfonamides, SPR- Spiramycin, Steroids- corticosteroids.

b. Painful adenopathy may respond to Indomethacin, prolonged adenopathy may respond to PYR/SULFA.

c. Infection during pregnancy as determined by seroconversion of the mother is treated with Spiramycin. If the fetus is confirmed to have toxoplasmosis by ultrasound, amniocentesis or cordocentesis then PYR/SULFA is given alternating with SPR.

d. Congenital toxoplasmosis is treated until the infant is 6 to 12 months old.

Acquired Toxoplasmosis During Pregnancy

Acutely infected pregnant women should be given 3 grams/day of Spiramycin divided three times a day once maternal infection is suspected or diagnosed in order to decrease transmission (McAuley et al., 1994). Spiramycin should be continued during pregnancy. Amniocentesis, fetal blood monitoring and fetal ultrasonography should be used to assess infection in the fetus. Ultrasonography should be done every 2 to 4 weeks as ventricular dilation may develop in as little as 10 days. Specific therapy should be initiated if fetal toxoplasmosis is diagnosed by either demonstration of fetal IgM, culture from amniotic fluid or fetal blood, PCR of amniotic fluid or fetal blood or ultrasonographic evidence of ventricular dilatation. Specific therapy consists of pyrimethamine (25 to 50 mg/day), sulfadiazine (3 to 4 gm/day) and folinic acid (5-15 mg/day) which should be administered to the mother (McAuley et al 1994). This treatment can be given alternatively with Spiramycin (3 gm/day) every three weeks until delivery. The majority of infants born to women treated with this regimen had subclinical disease at birth. Pyrimethamine should not be used in the first 14 - 16 weeks of pregnancy due to the possibility of teratogenicity.

Congenital Toxoplasmosis

At present a United States based clinical trial is in progress to address the optimal therapy and duration of therapy for children with congenital toxoplasmosis (R. McLeod, personal communication). Published results from this study suggest that treatment has benefit for congenital infection (McAuley et al., 1994). It is clear that even neonates who appear normal at birth (subclinical disease) may later demonstrate serious sequelae (primarily retinitis). Congenital toxoplasmosis is treated with a loading dose of pyrimethamine of (PYR) 2mg/kg/d for two days followed by PYR 1mg/kg/d or 15 mg/M^2 for 2 months and then PYR 15 mg/M^2 or 1 mg/kg every other day for the next 10 months. In addition, sulfadiazine or trisulfapyrimidine at 100 mg/kg/day in 2 divided doses and folinic acid 5 mg every other day is administered. Corticosteroids (1 mg/kg/d) should be added in patients with active macular disease or active CSF profiles (CSF protein $\geq$ 1 gm/dl).

Ocular Toxoplasmosis

There are limited trials of the different antitoxoplasmal drugs in this disorder (Holland et al 1996). Since the retinochoroiditis is self-limited, comparative trials are essential. Pyrimethamine and sulfadiazine appear to be effective in decreasing inflammation but do not appear to shorten the time course of the retinitis. Corticosteroids (Prednisone 1 to 2 mg/kg/day) are indicated if the macula, optic nerve head or papillomacular bundle are involved. When prednisone is given it is tapered when pigmentation (healing) begins. Clindamycin (1200 mg/d) has been used as an alternative drug, but it was inferior to pyrimethamine/sulfadizine (Tabbara and O'Connor, 1980).

Toxoplasma Infection in Immunocompromised Hosts

In AIDS patients treatment for *Toxoplasma* encephalitis is pyrimethamine 200 mg loading dose followed by 75-100 mg/day along with sulfadiazine 1 to 1.5 gm every 6 hours and folinic acid 10-25 mg/d (Liesenfeld et al., 1999). In patients intolerant of sulfadiazine, clindamycin 600 to 1200 mg every 6 hours can be used with pyrimethamine (Remington et al., 1991). Therapy is often started empirically and response is expected within 14 days. If no response is seen then brain biopsy is required for diagnosis. Corticosteroids are often used to control intracranial hypertension due to mass effect. Desensitization to sulfadiazine has been reported to be successful. Relapse of encephalitis occurs when treatment is stopped in about 30 percent of patients Relapse may take several weeks to become apparent (Renold et al., 1992). Patients are therefore maintained on pyrimethamine 25 to 75 mg/d, sulfadiazine 4gm/d, and folinic acid 10 mg/d after they have completed a 6 to 8 week course of primary treatment. HAART with restoration of immune function is also an important component of therapy. When the CD4+ is greater than 200, it may be possible to stop secondary prophylaxis.

Alternative combinations with preported efficacy in case reports include pyrimethamine with one of the following: clarithromycin 1 gm every 12 hours, or atovaquone 1250 mg every 12 hours, or azithromycin 1200 mg every day, or dapsone 100 mg every day.

DRUGS USED IN TREATMENT

Pyrimethamine (PYR)

Pyrimethamine (Daraprim), a substituted phenylpyrimidine, is an inhibitor of dihydrofolate reductase. Trimethoprim also inhibits this enzyme. Pyrimethamine is metabolized by the liver, readily absorbed from the gastrointestinal tract, and is found in the cerebrospinal fluid as well as other body compartments. The serum half-life of pyrimethamine is 35 to 175 hours and serum levels on a dose of 25 to 75 mg per day range from 1 to 4.5 mg/L (Weiss et al., 1988). Serum levels for an individual are not predictable due to the wide variation in absorption and serum half-life. CSF levels of pyrimethamine are 10 to 25 percent of the corresponding serum levels. Dose related bone marrow suppression with thrombocytopenia, neutropenia and anemia may develop. Folinic acid (leucovorin) is routinely given at a dose of 5 to 10 mg/d orally to prevent these effects. Folinic acid doses not inhibit the action of pyrimethamine on *T. gondii,* as this organism cannot take up folinic acid.

Sulfonamides (SULFA)

Sulfonamides inhibit dihydrofolic acid synthetase which is another enzyme involved in folic acid metabolism; thus they are synergistic with pyrimethamine. These drugs are well absorbed with good penetration into cerebrospinal fluid. Adverse reactions to sulfonamides are common, particularly in AIDS patients. Bone marrow suppression is seen and this responds to folinic acid. Hypersensitivity reactions with rash and Stevens-Johnson syndrome have been reported.

Macrolides and Lincomycins

Spiramycin (SPR) is a macrolide antibiotic available in many countries that has been used for the treatment of toxoplasmosis acquired during pregnancy in France. In the United States SPR is available by request from the U.S. Food and Drug Administration. Azithromycin, roxithromycin and clarithromycin are new macrolides that also have toxplasmocidal activity. It is believed that these compounds work by inhibition of the apicoplast. Clindamycin also is believed to work by inhibition of the apicoplast. The main side effect of clindamycin is diarrhea and pseudomembranous colitis.

It is quite clear that less toxic drugs for toxoplasmosis therapy in AIDS patients are needed. In addition, a drug active against the tissue cyst (bradyzoite) would make the radical cure of toxoplasmosis feasible and could be used as prophylaxis in seropositive immunocompromised patients.

PREVENTION

The following preventive measures can be taken to lesson the risk of acquiring *T. gondii* infection: (1) People should not eat raw or rare meat. This will prevent ingestion of viable tissue cysts, (2) People who prepare food should wash their hands and food preparation surfaces with warm soapy water after handling and preparing raw meat. This will inactivate tissue cysts, (3) Cat litter boxes should be changed regularly. Pregnant women and immunosuppressed individuals should not change the litter box. This will remove oocysts before they can become infective. (4) Gloves should be worn while gardening or hands should be washed after gardening or other contact with soil. This will prevent exposure to oocysts in the soil. (5). All fruits and vegetables should be washed before eating. This will help remove any oocysts that may be present on the surfaces of produce. *Toxoplasma gondii* infection can be prevented in cats by feeding them only commercially prepared cat food and keeping them inside to prevent hunting.

RECENT ADVANCES AND CONTEMPORARY CHALLENGES

The use of molecular genetic tools has greatly advanced our understanding of the functioning of the tachyzoite stage of *T. gondii*. Transfection techniques have been well developed for this pathogen (Soete et al., 1999), with reverse genetics proving to be a powerful tool for the elucidation of gene function. An EST project has been completed (http://www.cbil.upenn.edu/ParaDBs/) with ESTs from tachyzoites, bradyzoites and sporozoites. The genome of this organism is in progress with 1X coverage having been completed (http://ToxoDB.org/ToxoDB.shtml). The cell cycle of *T. gondii* has recently been described by Radke et al. (2001). Developmental biology studies of this organism are feasible due to the development of bradyzoite and tachyzoite markers and the ability of this organism to undergo differentiation *in vitro* (Weiss and Kim, 2000). Cell biology studies have defined many of the secretory pathways of this organism (Liendo and Joiner, 2000), as well as the assembly of the cytoskeleton and motility of this organism (Morrissette and Sibley, 2002). In addition, the recognition of the apicoplast (plastid) in *T. gondii* and other Apicomplexa has lead to a great deal of interest in plant-like metabolic pathways as therapeutic targets (Marechal and Cesbron-Delauw, 2001).

REFERENCES

Aramini, J. J., C. Stephen, J. P. Dubey, C. Engelstoft, H. Schwantje, and C. S. Ribble. 1999. Potential contamination of drinking water with *Toxoplasma gondii* oocysts. Epidemiology and Infection 122: 305-315.

Behymer, R. D., D. R. Harlow, D. E. Behymer, and C. E. Franti. 1973. Serologic diagnosis of toxoplasmosis and prevalence of *Toxoplasma gondii* antibodies in selected feline, canine, and human populations. Journal of the American Veterinary Medical Association 162: 959-963.

Bohne, W., J. Heesemann, and U. Gross. 1993. Coexistence of heterogenous populations of *Toxoplasma gondii* parasites within parasitophorous vacuoles of murine macrophages as revealed by bradyzoite specific monoclonal antibody. Parasitology Research 79: 485-487.

Bozzette, S., D. Findelstein, S. Spector , P. Frame, W. G. Powderly, W. He, L. Phillips, D. Craven, C. van der Horst, and Feinberg J.. 1995. A randomized trial of three antipneumocystis agents in patients with advanced human immunodeficiency virus infection. New England Journal of Medicine 332: 693-699.

Campagna, A. C. 1997. Pulmonary toxoplasmosis. Seminars in Respiratory Infection 12: 98-105.

Conley, F. K., K. A. Jenkins, and J. S. Remington. 1981. *Toxoplasma gondii* infection of the central nervous system: use of the peroxidase-antiperoxidase method to demonstrate *Toxoplasma* in formalin-fixed paraffin embedded tissue section. Human Pathology 12: 690-698.

Coutinho, S. G., R. Lobo, and G. Dutra. 1982. Isolation of *Toxoplasma* from the soil during an outbreak of toxoplasmosis in a rural area in Brazil. Journal of Parasitology 68: 866-868.

Couvreur, J., G. Desmonts, G. Tournieur, and M. A. Szusterkae. 1984. Etude d une serie homogene de 210 cases de toxplamose congenitale chez des nouurssons ages de 0 a 11 mois et depistes de facon prospective. Annals of Pediatrics (Paris) 31: 815-819.

Dannemann, B. R., W. C. Vaughan, P. Thulliez, and J. S. Remington. 1990. Differential aggulutination test for diagnosis of recently acquired infection with *Toxoplasma gondii*. Journal of Clinical Microbiology 28: 1928-1933.

Decavalas, G., M. Papapetropoulou, E. Giannoulaki, V. Tzigounis, and X. G. Kondakis. 1990. Prevalence of *Toxoplasma gondii* antibodies in gravidas and recently aborted women and study of risk factors. European Journal of Epidemiology 6: 223-226.

DiGiacomo, R. F., N. V. Harris, N. L. Huber, and M. K. Cooney. 1990. Animal exposures and antibodies to *Toxoplasma gondii* in a university population. American Journal of Epidemiology 131: 729-733

Dobrowolski, J. M., and L. D. Sibley. 1996. *Toxoplasma* invasion of mammalian cells is powered by the actin cytoskeleton of the parasite. Cell 84: 933-939.
Dubey, J. P. 1994. Toxoplasmosis. Journal of the American Veterinary Medical Association 205: 1593-1598.
_____.1998. *Toxoplasma gondii* oocyst survival under defined temperatures. Journal of Parasitology 84: 862-865
_____, and C. P. Beattie. 1988. *Toxoplasmosis of Animals and Man.* Boca Raton, CRC Press, pp. 1-40.
_____, and J. K. Frenkel. 1972. Cyst-induced toxoplasmosis in cats. Journal of Protozoology 19: 155-
177.
_____, and _____. 1976. Feline toxoplasmosis from acutely infected mice and the development of *Toxoplasma* cysts. Journal of Protozoology 23: 537-546.
_____, N. L. Miller, and J. K. Frenkel. 1970a.
Characterization of the new fecal form of *Toxoplasma gondii*. Journal of Parasitology
_____, _____, and _____.1970b. The *Toxoplasma gondii* oocyst from cat feces. Journal of Experimental Medicine 132: 636-662.
_____, M. C. Jenkins, D. W. Thayer, O. C. Kwok, and S. K. Shen.1996. Killing of *Toxoplasma gondii* oocysts by irradiation and protective immunity induced by vaccination with irradiated oocysts. Journal of Parasitology 82: 724-727.
_____, D. W. Thayer, C. A. Speer, and S. K. Shen. 1998. Effect of gamma irradiation on unsporulated and sporulated *Toxoplasma gondii* oocysts. International Journal for Parasitology 28: 369-375.
Fuentes, I., Rubio, J. M., C. Ramirez, and J. Alvar . 2001. Genotypiccharacterization of *Toxoplasma gondii* strains
associated with human toxoplasmosis in Spain:
direct analysis from clinical samples. Journal of Clinical
Microbiology 39: 1566-1570
Frenkel, J. K., A. Ruiz , and M. Chinchilla. 1975. Soil survival of *Toxoplasma* oocysts in Kansas and Costa Rica. American Journal of Tropical Medicine and Hygiene 24: 439-443.
_____, J. P. Dubey, and R. L. Hoff. 1976. Loss of stages after continuous passage of *Toxoplasma gondii* and *Besnoitia jellisoni*. Journal of Protozoology 23: 421-424.
_____, K. M. Hassanein, R. S. Hassanein, E. Brown, P. Thulliez, and R. Quintero-Nunez. 1995. Transmission of *Toxoplasma gondii* in Panama City, Panama: a five-year prospective cohort study of children, cats, rodents, birds, and soil. American Journal of Tropical Medicine and Hygiene 53: 458-468
Freyre, A., J. P. Dubey, D. D. Smith, and J. K. Frenkel 1989. Oocyst-induced *Toxoplasma gondii* infections in cats. Journal of Parasitology 75: 750-755.
Glasner, P. D., C. Silveira, D. Kruszon-Moran, M. C. Martins, M. Burnier, Jr., S. Silveira, M. E. Camargo, R. B. Nussenblatt, R. A. Kaslow, and R. Belfort Jr. 1992. An unusually high prevalence of ocular toxoplasmosis in Southern Brazil. American Journal of Ophthalmology 114: 136-144.
Grigg, M. E., and J. C. Boothroyd. 2001. Rapid identification of virulent type I strains of the protozoan pathogen *Toxoplasma gondii* by PCR-restriction fragment length polymorphism analysis at the B1 gene. Journal of Clinical Microbiology 39: 398-400.
Grover, C. M., P. Thulliez, J. S. Remington, and J. D. Boothroyd. 1990. Rapid prenatal diagnosis of congenital *Toxoplasma* infection by using polymerase chain reaction and amniotic fluid. Journal of Clinical Microbiology 28: 2297-2301.
Holliman, R. E., J. D. Johnsos, and D. Savva. 1990. Diagnosis of cerebral toxoplasmosis in association with AIDS using the polymerase chain reaction. Scandinavian Journal of Infectious Diseases 22: 243-244.

Holland, G. N., G. R. O'Connor, R. Belfort,Jr, and J. S. Remington. 1996. Toxoplasmosis. In: Pepose JS, Holland GN, Wilhelmus KR, eds. Ocular Infection and Immunity. St Louis: Mosby_Year Book. 1183-1223.
Howe, D. K. and L. D. Sibley. 1995. *Toxoplasma gondii* comprises three clonal lineages: correlation of parasite genotype with human disease. Journal of Infectious. Diseases 172: 1561-1566.
Isaac-Renton, J., W. R. Bowie, A. King, G. S. Irwin, C. S. Ong, C. P. Fung, M. O. Shokeir, and J. P. Dubey. 1998. Detection of *Toxoplasma gondii* oocysts in drinking water. Applied Environmental Microbiology 64: 2278-2280.
Israelski, D. M., and J. S. Remington. 1993. Toxoplasmosis in the non-AIDS immunocompromised host. In Remington J, Swarts M. eds. Current Clinical Toprics in Infectious Diseases vol. 13. London: Blackwell Scientific. Pp 322-356.
Jones, J. L., D. Kruszon-Moran, M. Wilson, G. McQuillan, T. Navin, and J. B. McAuley. 2001. *Toxoplasma gondii* infection in the United States: Seroprevalence and risk factors. American Journal of Epidemiology 154: 357-365.
Kapperud, G., P. A. Jenum, B. Stray-Pedersen, K. K. Melby, A. Eskild, and J. Eng. 1996. Risk factors for *Toxoplasma gondii* infection in pregnancy. Resultsof a prospective case-control study in Norway. American Journal of Epidemiology 144: 405-412.
Kniel, K. E. D. S. Lindsay, S. S. Sumner, C. R. Hackney, M. D. Pierson, and J.P. Dubey. 2002. Examination of survival of *Toxoplsama gondii* oocysts on raspberries and blue berries. Journal of Parasitology in press
Koppe, J. G., D. H. Loewer-Sieger, and H. De Roever-Bonnett. 1986. Results of 20 year follow up of congenital toxoplasmosis. American Journal of Opthalmology 101: 248-249.
Li, S., G. Maine, Y. Suzuki, F. G. Araujo, G. Galvan, J. S. Remington, and S. F. Parmley. 2000. Serodiagnosis of recently acquired *Toxoplasma gondii* infection with a recombinant antigen. Journal of Clinical Microbiology 38: 179-184.
Liesenfeld, O., C. Press, J. G. Montoya, R. Gill, J. L. Isaac-Renton, K. Hedman, and J. S. Remington. 1997. False-positive results in immunoglobulin M (IgM) *Toxoplasma* antibody tests and importance of confirmatory testing: the Platelia toxo IgM test. Journal of Clinical Microbiology 35: 174-178.
Liendo, A., and K. A. Joiner. 2000. *Toxoplasma gondii*: conserved protein machinery in an unusual secretory pathway? Microbes and Infection 2: 137-144
Liesenfeld O., S. Y. Wong, and J. S. Reminton. 1999. Toxoplasmosis in the setting of AIDS. In: Bartlett JG, Merigan TC, Bolognesi D. eds. Textbook of ADIS Medicine. Baltimore: Willimas and Wilkins. Pp. 225-259.
Lindsay, D. S., J. P. Dubey, B. L. Blagburn, and M. Toivio-Kinnucan. 1991. Examination of tissue cyst formation of *Toxoplasma gondii* in cell culturesusing bradyzoites, tachyzoites, and sporozoites. Journal of Parasitology 77: 126-132.
_____, M. A. Toivio-Kinnucan, and B. L. Blagburn. 1993. Ultrastructural demonstration of cystogenesis by various *Toxoplasma gondii* isolates in cell culture. Journal of Parasitology 79: 289-292.
_____, J. P. Dubey, J. M. Butler, and B. L. Blagburn. 1997. Mechanical transmission of *Toxoplasma gondii* oocysts by dogs. Veterinary Parasitology 73: 27-33.
_____, B. L. Blagburn, J. P. Dubey. 2002. Survival of nonsporulated *Toxoplasma gondii* oocysts under refrigerator conditions. Veterinary Parasitology 103: 309-313.
Luder, C. G., and F. Seeber. 2001. *Toxoplasma gondii* and MHC-restricted antigen presentation: on degradation, transport and modulation. International Journal of Parasitology 31: 1355-1369.
Marechal, E., and M. F. Cesbron-Delauw. 2001. The apicoplast: a new member of the plastid family. Trends in Plant Science 6: 200-5.
McAuley, J., K. M. Boyer, D. Patel, M. Mets, C. Swisher, N. Roizen, C. Wolters, L. Stein, M. Stein, and W. Schey. 1994. Early and longitudinal evaluations of treated infants and children

and untreated historical patients with congenital toxoplasmosis: The Chicago collaborative treatment trial. Clinical Infectious Diseases 18: 38-72.

McCabe, R. E., R. G. Brooks, R. F. Dorfman, and J. S. Remington. 1987. Clinical spectrum in 107 cases of toxoplasmic lymphadenopathy. Reviews of Infectious Diseases 9: 754-774.

McLeod, R., C. Brown, and D. Mack. 1994. Immunogenetics influence the outcome of *Toxoplasma gondii* infection. Research in Immunology 144: 61-65.

Mead, P. S., L. Slutsker, V. Dietz, L. F. McCaig, J. S. Bresee, C. Shapiro, P. M. Griffin, and R. V. Tauxe. 1999. Food-related illness and death in the United States. Emerging Infectious Diseases 5: 607-624.

Mitchell, C. D., S. S. Erlich, M. T. Mastrucci, S. C. Hutto, W. P. Parks,and G. B. Scott1990. Congential toxoplasmosis occurring in infants perinatally infected with human immunodeficiency virus 1. Pediatric Infectious Diseases Journal 9: 512-518.

Montoya , J. G., and J.S. Rermington. 1996. Toxoplasmic chorioretinitis in the setting of acute acquired toxoplasmosis. Clinical Infectious Diseases 23: 277-282.

_____, S. Parmley, O. Liesenfeld, O Jaffe and J. S. Remington JS. 1999. Use of the poluymerase chain reaction for diagnosis of ocular toxoplasmosis. Ophthalmology 106: 1554-1563.

Mordue, D. G., F. Monroy, M. La Regina, C. A. Dinarello, and L. D. Sibley. 2001. Acute toxoplasmosis leads to lethal overproduction of Th1 cytokines. Journal of Immunology 167: 4574-4584.

Morrissette, N. S., and L. D. Sibley. 2002. Cytoskeleton of apicomplexan parasites. Microbiology and Molecular Biology Reviews 66: 21-38.

Patronek, G. J. 1998. Free-roaming and feral cats--their impact on wildlife and human beings. Journal of the American Veterinary Medical Association 212: 218-226.

Naot, Y., D. R. Guptil, and J. S. Remington. 1982. Duration of IgM antibodies to *Toxoplasma gondii* after acute acquire toxoplasmosis. Journal of Infectious Diseases 145: 770.

Pfefferkorn, E. R., and L. C. Pfefferkorn. 1976. *Toxoplasma gondii*: Isolation and preliminary characterization of temperature-sensitive mutants. Experimental Parasitology 39: 365-376.

Radke, J. R., B. Striepen, M. N. Guerini, M. E. Jerome, D. S. Roos, and M. W. White. 2001. Defining the cell cycle for the tachyzoite stage of *Toxoplasma gondii*. Molecular and Biochemical Parasitology 115: 165-175.

Remington, J. S. 1974. Toxoplasmosis in the adult. Bulletin of the New York Academy of Medicine 50: 211-227.

_____, R. McLeod, and G. Desmonts. 1995. Toxoplasmosis, in Remington JS, Klein JO (eds): *Infectious Diseases of the Fetus and Newborn Infant*, 4th Edition, Philadelphia WB Saunders, pp 140-267.

_____, J. B. Wilder, F. Antunes, et al. 1991. Clindmycin for toxoplasmosis encephalitis in AIDS (letter). Lancet 338: 1142-1143.

Renold, C., A. Sugar, J. P. Chave, L. Perrin, J. Delavelle, G. Pizzolato, P. Burkhard,V. Gabriel and B. Hirschel. 1992. Toxoplasma encephalitis in patients with the acquired immunodeficiency syndrome. Medicine 71: 224-239.

Ross, R. D., L. A. Stec, J. C. Werner, M. S. Blumenkranz, L. Glazer, and G. A. Williams. 2001. Presumed acquired ocular toxoplasmosis in deer hunters. Retina 21: 226-229.

Ruiz, A., J. K. Frenkel, and L. Cerdas. 1973. Isolation of *Toxoplasma* from soil. Journal of Parasitology 59: 204-206.

Sabin, A. B. 1941. Toxoplasmic encephalitis in children. Journal of the American Medical Association 116: 801-807.

Sacks, J. J., D. G. Delgado, H. O. Lobel, and R. L. Parker.1983. Toxoplasmosis infection associated with eating undercooked venison. American Journal of Epidemiology 118: 832-838

Sengbusch, H. G., and L. A. Sengbusch. 1976. *Toxoplasma* antibody prevalence in veterinary personnel and a selected population not exposed to cats. American Journal of Epidemiology 103: 595-597

Sibley, L. D., and J. C. Boothroyd. 1992. Virulent strains of *Toxoplasma* comprise a single clonal lineage. Nature 359: 82-85.
____, D. G. Mordue, C. Su, P. M. Robben, and D. K. Howe. 2002. Genetic approaches to studying virulence and pathogenesis in *Toxoplasma gondii*. Philosophical Transactions of the Royal Society of London Biological Science 357: 81-88
Suzuki, Y. 2002. Host resistance in the brain against *Toxoplasma gondii*. Journal of Infectious Diseases 185S: S58-S65
Soete, M., B. Fortier, D. Camus, and J. F. Dubermatz. 1993. *Toxoplasma gondii*: Kinetics of bradyzoite- tachyzoite interconversion in vitro. Experimental Parasitology 76: 259-264.
_____, C. Hettman, and D. Soldati. 1999. The importance of reverse genetics in determining gene function in apicomplexan parasites. Parasitology 118: S53-S61
Sulzer, A. J., E. L. Franco, E. Takafuji, M. Benenson, K. W. Walls, and R. L. Greenup. 1986. An oocyst-transmitted outbreak of toxoplasmosis: patterns of immunoglobulin G and M over one year. American Journal of Tropical Medicine and Hygiene 35: 290-296.
Tabbara, K. F., and G. R. O'Connor. 1980. Treatment of ocular toxoplasmosis with clindamycin and sulfadiazine. Opthalmology. 87: 129-134.
Torres, R., M. Barr, M. Thorn, G. Gregory, S. Kiely, E. Chanin, C. Carlo, M. Martin, and J. Thornton. 1993. Randomized trial of dapsone and aerosolized pentamidine for the prophylaxis of *Pneumocystis carinii* pneumonia and toxoplasmic encephalitis. American Journal of Medicine 95: 573-583.
van Loon, A. M., J. T. M. van der Logt, F. W. A. Heessen, and J. van der Veen. 1983. Enzyme-linked immunosorbent assay that uses labeled antigen for detection of immunoglobulin M and A antibodies in toxoplasmosis: Comparison with indirect immunfluorescence and double-sandwich emzyme-linked immunosorbent assay. Journal of Clinical Microbiology 17: 997-1004.
Vogel, N., M. Kirisits, E. Michael, H. Bach, M. Hostetter, K. Boyer, R. Simpson, E. Holfels, J. Hopkins, D. Mack, M. B. Mets, C. N. Swisher, D. Patel, N. Roizen, L. Stein, M. Stein, S. Withers, E. Mui, C. Egwuagu, J. Remington, R. Dorfman, and R. McLeod. 1996. Congenital toxoplasmosis transmitted from an immunologically competent mother infected before conception. Clinical Infectious Diseases 23: 1055-1060.
Wallace, M. R., R. J. Rossetti, and P. E. Olson. 1993. Cats and toxoplasmosis risk in HIV-infected adults. Journal of the American Medical Association 269: 76-77.
Weiss, L. M., and D. Kim. 2000. The development and biology of bradyzoites of *Toxoplasma gondii*. Frontiers Bioscience 5: D391-D405.
_____, C. Harris, M. Berger, H. B. Tanowitz, and M. Wittner. 1988. Pyrimethamine concentration in serum and cerebrospinal fluid during treatment of *Toxoplasma* encephalitis in patients with AIDS. Journal of Infectious Diseases 157: 580-583.
_____, D. Laplace , P. M. Takvorian, H. B. Tanowitz, A. Cali, and M. Wittner. 1995. A cell culture system for study of the development of *Toxoplasma gondii* bradyzoites. Journal of Eukaryotic Microbiology 42: 150-157
Wilson, C. B., J. S. Reminsgton, S. Stagno, and D. W. Reynolds. 1980. Development of adverse sequela in children born with subclinical congenital *Toxoplasma* infection. Pediatrics 66: 767-774.
Wilson, M., J. S. Remington, C. Clavet, G. Varney, C. Press, and D. Ware. 1997. Evaluation of six commercial kits for detection of human immunoglubulin M antibodies to *Toxoplasma gondii*. Journal of Clinical Microbiology 35: 3112-3115.

BABESIOSIS

Paul Lantos and Peter J. Krause

Connecticut Children's Medical Center and The University of Connecticut School of Medicine

Etiologic Agents: Four species of babesia have been found to cause disease in humans: *Babesia microti, Babesia divergens,* WA-1, and MO-1.
Clinical Manifestations: Malaria-like symptoms of fever, fatigue, chills, sweats, headache, muscle aches, and anorexia. Severe cases may be fulminant and result in death.
Complications: Severe anemia, jaundice, acute respiratory failure, congestive heart failure, renal failure, coma, and death
Mode of Transmission: Ticks of the genus *Ixodes* are the primary vectors; transfusion and perinatal transmission occur in rare instances
Diagnosis: Identification of the parasite on thin blood smear or by PCR, antibody detection using IFA or Western blot
Treatment: Clindamycin and quinine or atovaquone and azithromycin; exchange transfusion in severe cases
Reservoir Hosts: Primarily white footed mice, also found in shrews, chipmunks, voles, and rats
Control Measures: Tick preventive measures

Babesiosis is an emerging zoonotic disease caused by tick-borne protozoa (Spielman et al., 1985; Krause, 2002; Dammin, 1978; Meldrum et al., 1992; Wilson, 1991; Krause et al., 1992). Although babesial species infect a wide variety of wild and domestic animals, the infection has long been recognized as an important disease in livestock. The first reference to babesiosis may have been in the Bible, where a widespread murrain or plague in cattle and other domestic animals is described in Exodus, Chapter 9, Verse 3: *"Behold, the hand of the Lord is upon thy cattle which is in the field, upon the horses, upon the asses, upon the camels, upon the oxen, and upon the sheep: there shall be a very grievous murrain."* The word "murrain" still is used to describe red-water fever, a form of babesiosis found in cattle in parts of Ireland (Dammin,1978). The pathogen was first described by the Hungarian microbiologist Babes who identified the microorganism in the blood of cattle in 1891. Five years later Smith and Kilbourne found that *Babesia bigemina*, the cause of Texas cattle fever, was transmitted by ticks. This seminal discovery was the first description of pathogens being transmitted by arthropods. The first human case of babesiosis was not described until 1957 when a Yugoslavian farmer died after becoming infected with *Babesia bovis*, another cattle babesia (Skrabalo and Deanovic, 1957). There are now more than 90 known species of *Babesia*, infecting wild and domestic mammals throughout the world (Spielman et al., 1985). The public health burden of human babesiosis is poorly described, but in certain endemic sites it may be as prevalent as gonorrhea.[8] Although there have been few pediatric cases of babesiosis, studies suggest that the infection is much more common in children than is currently diagnosed (Mathewson et al., 1984; Scimeca et al., 1986; Esernio-Jenssen et al.,1987; Wittner M et al., 1982).

EPIDEMIOLOGY

Babesial parasites infect a wide array of mammals. Some are host-specific, but others can infect multiple species. Humans are rare and incidental hosts of babesia that depend on other mammals for their life cycle. The species known to infect humans are *Babesia microti*, a babesia of mice, *Babesia divergens*, a babesia of cattle, and the recently described WA-1 and M-1 whose mammalian reservoirs are uncertain.

The most highly endemic area for babesiosis is the coastal region of the northeastern United States between Cape Cod and New Jersey (Spielman et al., 1985, Krause, 2002; Dammin, 1978). This includes the offshore islands of Nantucket, Martha's Vineyard, Block Island, and eastern Long Island. Human disease caused by *B. microti*, which is responsible for the cases in this region, has been reported in Connecticut, Rhode Island, Massachusetts, New York, and New Jersey. *Babesia microti* also is endemic in the upper Midwest, including Minnesota and Wisconsin (Steketee et al., 1985). The incidence of *B. microti* infections has risen sharply during the last two decades (Krause, 2002; Meldrum et al., 1992, Dammin et al., 1981). A related species, designated WA-1 has been identified in cases from the western United States (Persing et al., 1995). Another organism called MO-1, which is related to *B. divergens*, was the causative agent in a single case from Missouri (Herwaldt et al., 1996). Other North American cases of babesiosis have been reported from the southeastern United States and from Mexico.

Approximately 30 cases of babesiosis have been identified in Europe. *Babesia divergens* was the known pathogen in more than 75 percent of these cases and more than half were from France, Ireland, and the United Kingdom (Garnham, 1980; Brasseur and Gorenflot, 1996). Additional case reports have arisen from Germany, the former Yugoslavia, and the former Soviet Union. *Babesia divergens* is a babesia of cattle and is transmitted by the *Ixodes ricinus* tick. Recent studies suggest that human *B. microti* infection may be endemic in Europe as well (Foppa et al., in press).

Babesiosis also has been described in Asia and Africa (Shih et al., 1997; Bush et al., 1990; Hsu and Cross, 1977). An isolated case of babesiosis was reported in Taiwan (Shih et al., 1997). This infection was caused by a babesia related to *B. microti*, and tentatively identified as TW-1. Surveys of rodents in the area revealed an 83 percent seroprevalence. Babesiosis is an important veterinary pathogen in African livestock but human cases are very rare (Jack and Ward, 1980). The relative lack of cases from tropical locations may be explained by cross-immunity from other parasites and misdiagnosis of babesiosis as malaria.

Transmission

Babesiosis is transmitted by small, hard-bodied ticks in the genus *Ixodes* and rarely through blood transfusion or transplacentally (Spielman et al., 1985; Esernio et al., 1987; Wittner et al., 1982). In the northeastern United States, the deer tick *I. scapularis* (also known as *I. dammini*) is the vector for *B. microti*. This tick also transmits *Borrelia burgdorferi* and *Anaplasma phagocytophila*, the causative agents of Lyme disease and human granulocytic ehrlichiosis, respectively (Spielman et al., 1985).

Ixodid ticks have three active stages in their life cycle: larval, nymphal, and adult, all of which require a blood meal from a mammalian host (Figure 1) (Spielman et al., 1985). Babesia may be ingested during the larval or nymphal stages and persist through the molt to the next developmental stage. Babesia infect the salivary glands of the tick and are transmitted to the host during feeding. The nymph, which feeds during late spring and summer, is the most common vector for human infection.

The white-footed mouse, *Peromyscus leucopus*, (Figure 2) is the primary reservoir for *B. microti*, although the parasite also has been found in shrews, chipmunks, voles, and rats (Spielman et al., 1985). As many as two thirds of white-footed mice are parasitemic in highly endemic areas. While rodents comprise the principal reservoir of *B. microti*, the white-tailed deer (*Odocoileus virginianus*) is also critical to the ecology and life cycle of babesiosis. Adult ticks preferentially feed on white-tailed deer and it is during this blood meal that mating occurs. While deer

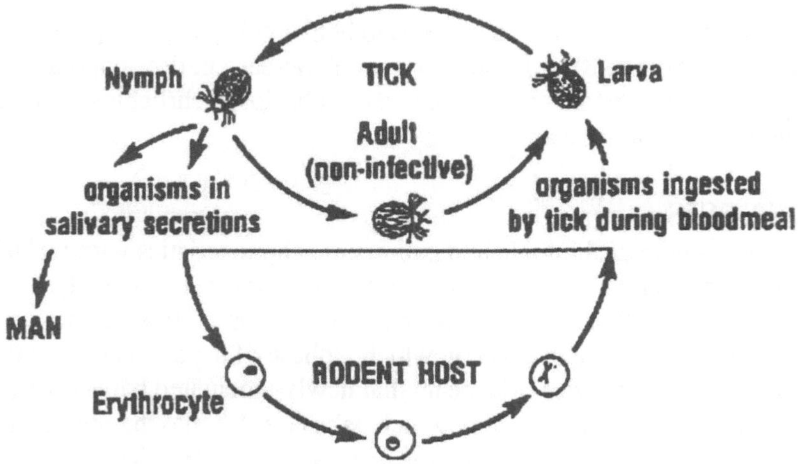

Figure 78–1. Life cycle of *Babesia microti*.

Figure 1a. Life cycle of Babesia microti, *modified from Ruebush (1984).*

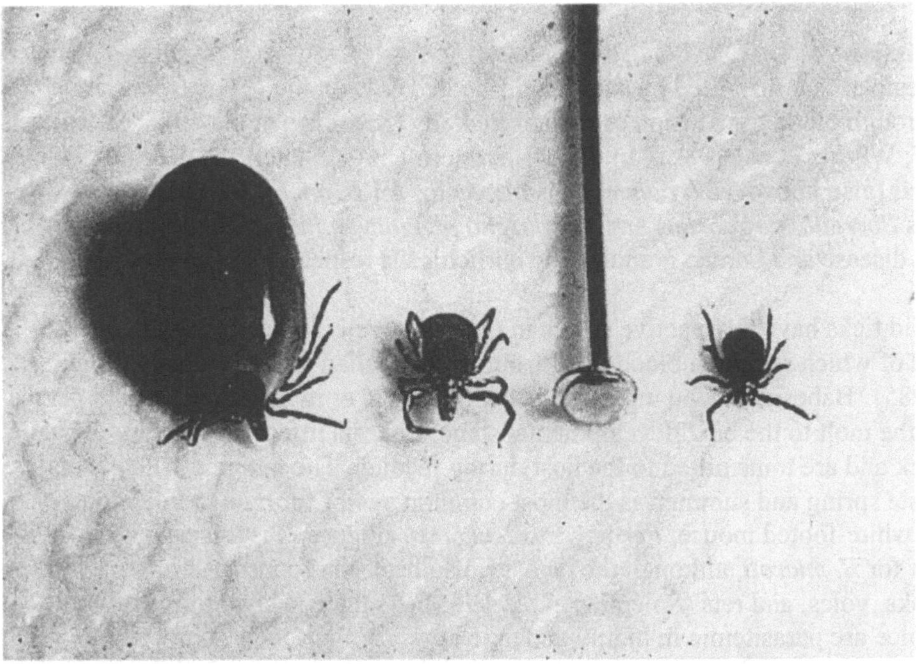

Figure 1b. Ixodes scapularis *ticks and a common pin. The ticks are, left to right, an engorged female, an adult female, and an adult male. Courtesy Mike Frigione, Pfizer, Inc.*

do not become infected by *B. microti*, they serve to both amplify the tick population and expand its range (Figure 3). It is the exponential increase in deer populations that is thought responsible for the sharp rise in cases of babesiosis, ehrlichiosis, and Lyme disease during the past quarter century.

PATHOGENESIS

Babesia share many microbiologic and pathogenic characteristics with malaria parasites. Like plasmodia, babesia are intraerythrocytic parasites (Figure 4). Sporozoites are introduced directly into the host bloodstream by tick saliva (Spielman et al., 1985). Unlike malaria, in which a phase of hepatic parasitism precedes infection of erythrocytes, it appears that newly inoculated babesia directly infect red blood cells. The mechanism by which babesia enter into the red cell is poorly understood, but it is thought to involve components of the alternative complement pathway including the C3b receptor (Jack and Ward, 1980).

After entry, the parasite buds asexually to form either two or four daughter cells. Tetrads of merozoites may form the rarely seen "Maltese crosses" that are

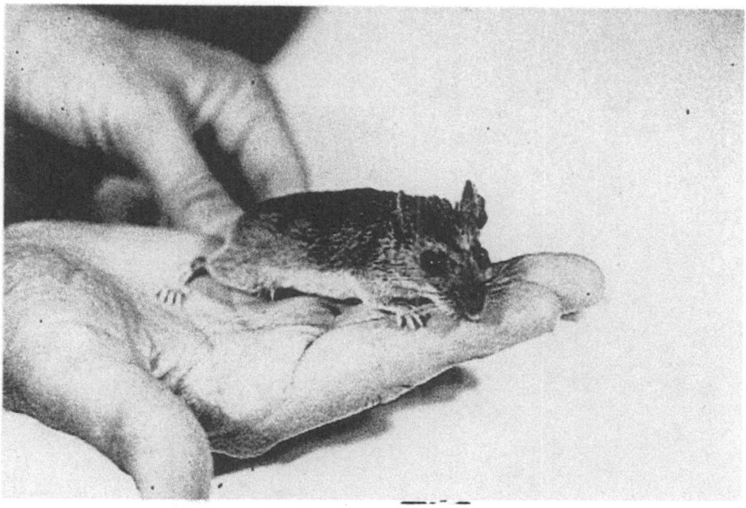

Figure 2. White-footed mouse.

Figure 3. White-tailed deer.

pathognomonic for babesiosis. The red cell membrane is ultimately lysed during trophozoite release. These go on to infect other red cells, creating a cycle of infection and hemolysis. Hemolysis in babesial infections is an asynchronous process in contrast to the synchronized and massive episodes of hemolysis seen in malaria. This distinction helps to explain why babesial infections are not associated with the discrete paroxysms of fever and rigors that characterize malaria.

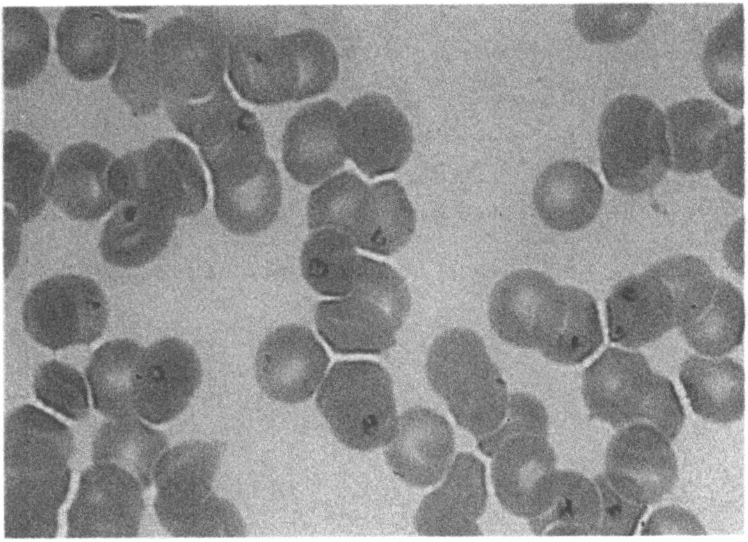

Figure 4. Ring forms of B. microti *in a human blood film.*

The protean symptoms of babesiosis are attributable to hemolysis and the immune response elicited by parasitemia (Homer et al., 2000). Hemolytic anemia occurs with the corresponding findings of hyperbilirubinemia, hemoglobinuria, decreased haptoglobin, and reticulocytosis. Infected and deformed erythrocytes also can cause vasoocclusive symptoms, including tissue ischemia and necrosis, hepatosplenomegaly, and hepatic injury. Systemic symptoms, such as fever, hypotension, and the systemic inflammatory response are mediated by cytokines. Among those implicated in babesiosis are complement, kallikrein, immune complexes, tumor necrosis factor, and fibronectin. Rodent studies have demonstrated that WA-1 infection produces pulmonary endothelial changes, inflammation, and capillary leakage, chiefly mediated by CD8+ T cells, that result in pulmonary edema and respiratory distress (Hemmer et al., 1999). Similar changes were not seen in *B. microti* infections and may help explain the more severe disease associated with WA-1 infection.

Asplenia is the most important risk factor for severe babesiosis, emphasizing the central role of the reticuloendothelial system in this infection (Rosner et al., 1984). The spleen coordinates the removal of infected erythrocytes from the circulation, the ingestion of free parasites by mononuclear phagocytes, and the generation of antibabesial antibodies. Human and animal studies have identified many components of the antibabesial immune response, involving both cellular and humoral mechanisms. These include T and B lymphocytes, polymorphonuclear leukocytes, complement, immunoglobulins, and inflammatory cytokines such as tumor necrosis factor, gamma-interferon, and interleukin-2 (Homer et al., 2000). While cytotoxic CD8+ T cells predominate over CD4+ T cells during acute babesiosis, sustained immunity appears to depend on CD4+ T cells and gamma-interferon (Homer et al., 2000; Clawson et al., in press). Immunity is often incomplete. Parasitemia may last months and reinfection can occur (Krause et al., 1998). Antigenic variation among babesia has been demonstrated and may help explain these findings.

CLINICAL MANIFESTATIONS

Nonspecific, flu-like symptoms characterize the clinical manifestations of babesiosis (Krause, 2002; Krause et al., 1996a; Ruebush et al., 1977a, Ruebush et al., 1977b). One to six weeks following an infectious tick bite, patients gradually develop fever, malaise, and fatigue (Table 1). Chills, sweats, headache, myalgias, and arthralgias are the most commonly reported symptoms (Krause, 2002).

Table 1. Symptoms of babesiosis

Symptom	Outpatient (n=41)	Inpatient (n=173)	Total (n=214)
Fever	68	89	85
Fatigue	78	79	79
Chills	39	68	63
Sweats	41	56	53
Headache	75	32	39
Myalgia	37	32	33
Anorexia	25	24	24
Cough	17	23	22
Arthralgia	31	17	18
Nausea	22	9	16

Outpatient cases are from Ruebush et al. (1977) and Krause et al. (1996).
Inpatient cases are from Meldrum et al. (1992) and Hatcher et al. (2001).

Gastrointestinal manifestations including abdominal pain, anorexia, nausea, vomiting, and weight loss may occur. Upper respiratory symptoms include sore throat, conjunctival injection, and nonproductive cough. Neurologic manifestations, including photophobia, hyperesthesia, depression, and emotional lability have been less frequently reported. Abnormal physical signs are infrequent and generally non-specific (Rosner et al., 1984; Ruebush et al., 1977b, Krause et al., 1996a). Fever is the most frequent physical finding. Splenomegaly, pallor, and jaundice are noted in cases with moderate to severe hemolysis. Conjunctivitis, pharyngitis, and retinal hemorrhages and infarcts may occur. Rash is uncommon but petechiae and bruising may be seen.

Fulminant illness resulting in prolonged convalescence or death has been reported in about five percent of cases of *B. microti* infection (Meldrum et al., 1992; Hatcher et al., 2001). Patients at greatest risk of severe disease include those who are asplenic, suffering from a malignancy, HIV, or a primary immunodeficiency, or receiving immunosuppressive drugs (Wittner et al., 1982; Rosner et al., 1984; Falgas and Klempner, 1995). Patients over the age of 50 also are predisposed to more serious infection. Disease caused by *B. divergens* appears to be more severe than the typical case of *B. microti* (Garnham, 1980). The mortality in severe cases of *B. divergens* infection exceeds 50%, although most of these have been reported in asplenic individuals (Brasseur and Gorenflot, 1996). Patients experiencing fulminant babesiosis frequently develop pulmonary edema or the adult respiratory distress syndrome (Table 2) (Hatcher et al., 2001). They also may develop high fever, severe hemolytic anemia, coagulopathy, jaundice, congestive heart failure, renal failure, and coma. While long term complications are uncommon, parasitemia may persist for months to years in some patients with subsequent recrudescence of disease (Krause et al., 1998).

Table 2. Complications of babesiosis in 34 consecutive hospitalized patients

Complication	Frequency (%)
Acute respiratory failure	21
DIC	18
Congestive heart failure	12
Coma/lethargy	9
Renal failure	6
Death	9

The mean age of patients was 53 years, median age 43 years, range 3 months to 85 years . From Hatcher et al. (2001)

Subclinical cases are common in endemic areas, especially among children. While only a few of the more than 300 reports of babesiosis are of infected children, seroprevalence for *B. microti* is similar in children and adults (Figure 5) (Krause et al., 1992). Although babesiosis in children generally is milder than that suffered by adults, three of eight neonates reported with *B. microti* infection experienced moderate to severe disease (Scimeca et al., 1986; Esernio-Jenssen, 1987; Dobroszycki et al., 1999). Clinical manifestations included fever, irritability, splenomegaly, and parasitemia of five to eight percent. Each of these infants was treated with intravenous clindamycin and quinine as well as a blood transfusion for anemia.

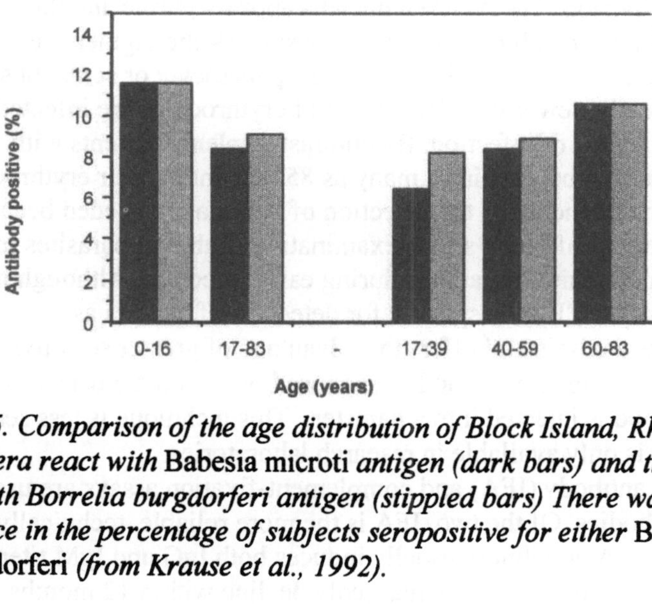

Figure 5. Comparison of the age distribution of Block Island, Rhode Island residents whose sera react with Babesia microti *antigen (dark bars) and those whose sera react with Borrelia burgdorferi antigen (stippled bars) There was no significant difference in the percentage of subjects seropositive for either* B. microti *or* B. burgdorferi *(from Krause et al., 1992).*

Babesia microti and *Borrelia burgdorferi* (the causative agent of Lyme disease) share the same mouse reservoir and tick vector. In areas endemic for both pathogens, between 10 and 20 percent of patients experiencing Lyme disease are concurrently infected with the babesial parasite (Krause et al., 1996a). Although such patients suffer a greater number of acute symptoms that persist longer than patients experiencing Lyme disease alone, the rate of joint, neurologic, and cardiac complications may be similar.

DIAGNOSIS

Patients experiencing babesiosis often have laboratory abnormalities that reflect hemolysis. These include anemia, reticulocytosis, decreased haptoglobin, hemoglobinuria, and elevations of lactate dehydrogenase, liver transaminases, alkaline phosphatase, and unconjugated bilirubin (Krause, 2002). A complete blood count is a useful screening study since anemia and thrombocytopenia are commonly observed and parasites may be seen on the blood smear. The leukocyte count may b normal or decreased. Some patients have an elevated erythrocyte sedimentation rate Marked elevations in elevated blood urea nitrogen occur in severe cases.

Laboratory methods that are used to make a specific diagnosis of babesiosis include microscopic examination of blood smears, identification of babesial antibody, polymerase chain reaction (PCR), and small animal inoculation. Intraerythrocytic organisms may be identified on Giemsa stained smears of peripheral blood (Healy and Ruebush, 1980). Babesia are 1 to 5 μm in diameter and may be spherical, oval, or teardrop-shaped. The most common morphology is the ring form, which closely resembles the classic "signet rings" seen in malaria. A pathognomonic yet rarely seen form is the Maltese cross, or tetrad, which occurs after two cycles of cell division. The babesial nucleus appears purple and the cytoplasm blue in Giemsa stained blood smears. Babesia lack the pigment seen in some forms of *Plasmodium* and do not display visible gametocyte or schizont stages as seen in malaria. Typically, fewer than 10 percent of erythrocytes are infected, particularly early in the course of infection. By contrast, asplenic patients with fulminant disease may harbor babesia in as many as 85 percent of their erythrocytes

Additional diagnostic methods for the detection of babesia are needed because c the relatively poor sensitivity of blood smear examination. Babesial parasites remair undetected in more than one third of patients during early infection. Although more time consuming and costly, PCR is as specific for detection of babesia as microscopic blood smear analysis and offers the advantage of greater sensitivity (Krause et al., 1996b). Babesiosis also can be diagnosed by xenodiagnosis in which sample of a patient's blood is injected into a hamster. This technique is less sensiti than PCR and generally is only available in research laboratories.

Immunofluorescent antibody (IFA) and complement-fixation assays are used to detect anti-babesial antibodies. Of the two, IFA is the more reliable, technically simpler, and inexpensive. Acute illness usually induces both IgG and IgM titers tha exceed 1:1024, although these levels often markedly decline within 12 months. Son cross-reactivity to *Plasmodium* occurs, but the titers generally are 1:16 or lower in patients experiencing malarial infection. Both immunoblot and enzyme-linked immunosorbent assays for detection of *B. microti* antibodies have been recently described (Ryan et al., 1999).

TREATMENT AND PREVENTION

The treatment of choice for babesiosis is a 7 to 10 day course of clindamycin and quinine (Table 3) (Centers for Disease Control, 1983). This combination was first used successfully in an 8-week-old infant with transfusion-associated babesios that was initially thought to be malaria. Since then it has been found to provide rap clinical and microbiologic improvement in both adults and children. Recently, the combination of azithromycin and atovaquone were found to be equally effective in treating adults experiencing babesiosis (Krause et al., 2000; Bonan et al., 1998). Th combination was associated with fewer adverse reactions than clindamycin and quinine (Krause et al., 2000). Although azithromycin and atovaquone each have been safely prescribed for children for other infectious conditions, the combination has not been tested in children. Clindamycin and quinine remain the combination of choice for treatment of pediatric babesiosis.

Table 3. Treatment of babesiosis

ANTIBIOTICS

Antibiotic	**Dose**		**Frequency**
Clindamycin	Adult:	600 mg	Every 6 hours
	Child:	5 mg/kg	
and			
Quinine	Adult:	650 mg	Every 6 hours
	Child:	25 mg/kg	
or			
Atovaquone	Adult:	750 mg	Every 12 hours
and			
Azithromycin	Adult:	500 to 600 mg	On day 1
		250 to 600 mg	On subsequent days

All antibiotics are administered for 7 to 10 days

EXCHANGE TRANSFUSION

Exchange transfusion or partial exchange transfusion should be considered for treatment of severe cases of babesiosis.

For patients who fail to respond to either clindamycin and quinine or atovaquone and azithromycin, pentamidine and trimethoprim-sulfamethoxazole might be considered since this combination was reported to be effective in one case of infection with *B. divergens* (Raoult et al., 1987). The combination of azithromycin and quinine was effective in a Taiwanese case after failure of clindamycin and quinine (Shih et al., 1998). Other antiprotozoal drugs are not useful against *Babesia*. Chloroquine offers some symptomatic relief but has inadequate antiparasitic activity. Quinacrine, primaquine, pyrimethamine, pyrimethamine-sulfadoxine, sulfadiazine, and tetracycline are all ineffective. Finally, exchange transfusion should be considered in patients with severe babesiosis (Bonoan et al., 1998; Jacoby et al., 1980). Even partial exchange transfusions may significantly reduce the parasite burden and, in combination with effective antibabesial chemotherapy, may avert mortality.

Measures to prevent babesiosis are the same as those recommended for the prevention of other tick-born infections such as Lyme disease (Spielman et al., 1985). Preventative measures are advisable for people in endemic areas with frequent exposure to ticks. Both people and pets should be examined for ticks, particularly after contact with foliage or tall grasses. Any attached tick should be removed promptly by grasping its mouth parts with forceps. Long pants tucked into socks will minimize the risk of exposure to skin. Clothing can be impregnated with insecticide agents, such as permethrin, dimethyl phthalate, or diethyltoluamide. These measures are particularly important in immunocompromised individuals at risk of severe disease.

Environmental interventions aimed at reducing deer, mouse, or tick populations have not been as successful as individual preventive measures (Spielman et al., 1985). While effective vaccines exist for bovine babesiosis caused by *B. bovis* and *B. bigemina*, no such vaccine exists for species that infect humans. The American Red Cross excludes blood donors with a history of babesiosis, and it has been recommended that persons from endemic areas who have a history of recent fever be excluded as well.

RECENT ADVANCES AND CONTEMPORARY CHALLENGES

The epidemiology of babesiosis both in the United Sates and worldwide is poorly understood. Current studies suggest that the incidence of human infection is markedly underestimated. The pathogenesis of the increased risk of serious disease experienced by persons over 50 years of age and those simultaneously infected by the Lyme spirochete are unclear. The pathologic events that contribute to life-threatening disease have not been elucidated either at the cellular or molecular level. Development of new standardized diagnostic techniques is needed such as an ELISA antibody test using recombinant antigen. Diagnostic screening of donated blood in endemic areas for evidence of babesial infection might be useful in decreasing the risk of transfusion transmitted babesiosis. Although a vaccine for cattle babesia has been developed, no human vaccine is currently available.

REFERENCES

Bonoan JT, Johnson DH, Cunha BA. 1998. Life- threatening babesiosis in an asplenic patient treated with exchange transfusion, azithromycin, and atovaquone. *Heart & Lung* 27: 424-428.

Brasseur P, Gorenflot A. 1996. Human babesiosis in Europe. *Roc Akadem Med W Bialym* 41: 117-22.

Bush JB, Isaacson M, Mohamed AS. 1990. Human babesiosis: a preliminary report of two suspect cases in southern Africa. *S Afr J Med* 78:699.

Centers for Disease Control. 1983. Clindamycin and quinine treatment for *Babesia microti* infections. *MMWR* 32: 65-72.

Clawson M, Paciorkowski N, Rajan TV, et al. Cellular immunity, but not IFN-Y, is essential for resolution of *Babesia microti* infection in BALB/c mice. *Infect Immun* (in press).

Dammin, GJ, Spielman A, Benach JL, et al. 1981. The rising incidence of clinical *Babesia microti* infection. *Hum Pathol* 12:398-400.
Dammin GJ. Babesiosis. 1978. *In* Weinstein L, Fields BN (eds.): *Seminars in Infectious Disease*, New York, Stratton, 169-199.
Dobroszycki J, Herwaldt BL, Boctor F, et al. 1999. A cluster of transfusion-associated babesiosis cases traced to a single asymptomatic donor. *JAMA* 281: 927-930.
Esernio-Jenssen D, Scimeca PG, Benach JL, Tenenbaum JJ. 1987. Transplacental/perinatal babesiosis. *J Pediatr* 110: 570-572.
Falagas ME, Klempner MS. 1995. Babesiosis in patients with AIDS: A chronic infection presenting as fever of unknown origin. *Clin Inf Dis* 22: 809-812.
Foppa I, Krause PJ, Spielman A, Goethert H, Gern L, Brand B, Telford S. Entomologic and serologic evidence of zoonotic transmission of *Babesia microti* in Eastern Switzerland. *Emerg Infect Dis* (in press)
Garnham, PCC. 1980. Human babesiosis: European aspects. *Trans R Soc Trop Med Hyg* 74:153-155. Herwaldt BL, Persing DH, Precigout EA, et al. 1996. A fatal case of babesiosis in Missouri: identification of another piroplasm that infects humans. *Ann Intern Med* 124: 643-650.
Krause PJ. 2002. Babesiosis. *Med Clin N Amer* 86: 361-374.
Hatcher JC, Greenberg PD, Antique J, et al. 2001. Severe babesiosis in Long Island: Review of 34 cases and their complications. *Clin Infect Dis* 32: 1117-1125.
Healy GR, Ruebush TK, II. 1980. Morphology of *Babesia microti* in human blood smears. *Am J Clin Pathol* 73:107-109.
Hemmer RM, Wozniak EJ, Lowenstine LJ, et al. 1999. Endothelial cell changes are associated with pulmonary edema and respiratory distress in mice infected with the WA1 human *Babesia* parasite. *J Parasitol* 85: 479-489.
Homer MJ, Aguilar-Delfin I, Telford SR, Krause PJ, Persing DH. Babesiosis. 2000. *J Microbiology Reviews* 13: 451-469.
Hsu NH, Cross JH. 1977. Serologic survey for human babesiosis on Taiwan. *J Formosan Med Assoc* 76:950-954.
Jack RM, Ward PA. 1980. *Babesia rodhaini* interactions with complement: Relationship to parasitic entry into red cells. *J Immunol* 124: 1566-1573.
Jacoby GA, Hunt JV, Kosinski KS, et al. 1980. Treatment of transfusion-transmitted babesiosis by exchange transfusion. *N Engl J Med* 303:1098-1100.
Krause PJ, Lepore T, Sikand VK, et al. 2000. Atovaquone and azithromycin for the treatment of human babesiosis. *N Engl J Med* 343: 1454-1458.
Krause PJ, Spielman A ,Telford S, et al. 1998. Persistent parasitemia following acute babesiosis. *New Engl J Med* 339: 160-165.
Krause, PJ, Telford, SR, Spielman, A, et al. 1996a. Concurrent Lyme disease and babesiosis: evidence for increased severity and duration of illness. *JAMA* 275: 1657-1660.
Krause, P. J., Telford, S. R., Spielman, A., et al. 1996b. Comparison of PCR with blood smear and inoculation of small animals for diagnosis of *Babesia microti* parasitemia. *J Clin Microbiol* 34: 2791-2794.
Krause PJ, Telford SR III, Pollack RJ, et al. 1992. Babesiosis: an underdiagnosed disease of children. *Pediatrics* 89: 1045-1048.
Mathewson HO, Anderson AE, Hazard GW. 1984. Self-limited babesiosis in a splenectomized child. *Ped Infect Dis* 3:148-149.
Meldrum SC, Birkhead GS, White DJ, et al. 1992. Human babesiosis in New York State: An epidemiologic description of 136 cases. *Clin Inf Dis* 15: 1019-1023.
Persing, D. H., Herwaldt, B. L., Glaser C., et al. 1995. Infection with a *Babesia*-like organism in northern California. *N Engl J Med* 332: 298-303,.
Raoult D, Soulayrol L, Toga B, et al. 1987. Babesiosis, pentamidine, and cotrimoxazole. *Ann Intern Med* 107: 944.

Rosner F, Zarrabi MH, Benach JL, Habicht GS. 1984. Babesiosis in splenectomized adults: Review of 22 reported cases. *Am J Med* 76: 696-701.
Ruebush TK, II. 1984. Babesiosis. In: Strickland GT (Ed.): Hunter's *Tropical Medicine*. Philadelphia, PA: W.B. Saunders.
Ruebush TK, II, Juranek DD, Chisholm ES, et al. 1977a. Human babesiosis on Nantucket Island: Evidence for self-limited and subclinical infections. *N Engl J Med* 297: 825-827.
Ruebush, TK, Cassaday PB, Marsh HJ, et al. 1977b. Human babesiosis on Nantucket Island: Clinical features. *Ann Intern Med* 86: 6-9.
Ryan R, Krause PJ, Radolf J, et al. 2001. *Clin and Diag Lan Immunol* 8: 1177-1180.
Scimeca PG, Weinblatt ME, Schonfeld G, Kaplan MH, Kochen JH. 1986. Babesiosis in two infants from Eastern Long Island, N.Y. *Am J Dis Child* 140: 971.
Shih CM et al. 1997. Human babesiosis in Taiwan: asymptomatic infection with a *Babesia microti*-like organism in a Taiwanese woman. *J Clin Microbiol* 85: 450-454.
Shih CM et al. 1998. Ability of azithromycin in combination with quinine for the elimination of babesial infection in humans. *Am J Trop Med Hyg* 59: 509-512.
Skrabalo Z, Deanovic Z.1957. Piroplasmosis in man. *Doc Med Geogr Trop* 9: 11-16.
Spielman A, Wilson ML, Levine JF, et al. 1985. Ecology of *Ixodes dammini*-borne human babesiosis and Lyme disease. *Ann Rev Entomol* 30: 439-460.
Steketee, RW, Eckman, MR, Burgess, EC, et al. 1985. Babesiosis in Wisconsin: A new focus of disease transmission. *JAMA* 253: 2675-2678.
Wilson ME. 1991. A World Guide to Infections: Diseases, Distribution, Diagnosis. New York, Oxford University Press.
Wittner M, Rowin KS, Tanowitz HB, et al. 1982. Successful chemotherapy of transfusion babesiosis. *Ann Intern Med* 96: 601-604.

OTHER NOTEWORTHY ZOONOTIC PROTOZOA

Oscar J. Pung

Department of Biology, Georgia Southern University, Statesboro, Georgia

AMERICAN CUTANEOUS LEISHMANIASIS

Etiologic Agent: *Leishmania mexicana*
Clinical Manifestations: Benign, often self-limiting cutaneous lesions are typical. These usually appear in the form of papules or nodules that may eventually ulcerate. Lesions are found at sites of sandfly bites, often on the face and ears.
Complications: Disseminated cutaneous lesions and lesions involving the nasal mucosa have been reported.
Mode of Transmission: The parasite is transmitted by the bite of infected sandflies belonging to the genus *Lutzomyia.*
Diagnosis: Detection of the amastigote stages of the parasite in dermal macrophages from stained sections of tissue or scraping from the edge of ulcers is the most common method of diagnosis. Promastigotes can be detected in cultured lesional tissues.
Drug(s) of Choice: Cutaneous lesions often heal without treatment. In cases where they do not, pentavalent antimony (Pentostan) has been used successfully. Local heat, cryotherapy, and surgical excision are also effective in some instances.
Reservoir Hosts: Southern plains woodrat, white-throated woodrat
Control Measures: Destruction of woodrat habitat and nests near human dwellings to reduce reservoir host populations. The use of fine mesh window screens, residual insecticides in the home, and insect repellents outdoors may decrease the risk of sandfly bites.

Etiology/Natural History

The *Leishmania* parasites are found on nearly every continent and are responsible for cutaneous, mucocutaneous, and visceral illnesses ranging in severity from benign, self-limiting skin lesions to fatal systemic disease. Autochthonous leishmania infection of persons in the U.S.A. is rare. Only a few dozen cases of localized cutaneous leishmaniasis due to *L. mexicana* infection have been reported from southern and central Texas since 1903 (McHugh et al., 1996). Most cases of leishmaniasis diagnosed in the U.S.A. are acquired in other endemic countries. It is worthwhile noting that *Leishmania infantum*, an agent of human visceral leishmaniasis in the Eastern Hemisphere, infects a significant percentage of foxhounds in the U.S.A. and Canada (Enserink, 2000). As a result, there is concern that *L.*

infantum is now endemic in North America though no human cases have been reported to date.

Leishmania parasites are transmitted by an infected female sandfly when it takes a blood meal from an infected reservoir or human and ingests a nonflagellated, intracellular stage of the parasite called an amastigote (Leishman-Donovan body). The amastigote transforms into a flagellated promastigote in the sandfly, migrates to the gut of the fly, multiplies, and migrates to the salivary gland. Infective metacyclic promastigotes enter a new host when the fly takes its next blood meal (Despomier et al., 2000; Garcia et al., 1997). In the mammalian host, the promastigote is phagocytized by cutaneous macrophages or dendritic cells in which it transforms into an amastigote and multiplies in the phagolysosome. The cell is eventually disrupted and the newly released parasites are taken up by other neighboring phagocytic cells or are ingested by a sandfly (Solbach and Laskay, 2000).

Epidemiology

Thousands of people in Mexico and Central America suffer from cutaneous leishmaniasis due to *L. mexicana*. *Leishmania mexicana* infection in these localities is an occupational hazard, primarily affecting men who work in tropical forests where several species of sylvatic rodents serve as reservoir hosts for the parasite. In fact, the disease is traditionally known as chiclero ulcer because it is common among laborers who harvest chicle tree gum (Roberts and Janovy, 1996).

The epidemiology of *L. mexicana* infection in southern Texas differs in that it is neither sylvatic nor occupationally acquired (McHugh et al., 1996). Infected patients typically reside in rural areas or on the edge of suburban neighborhoods in close proximity to semi-arid brush country dominated by mesquite brush and prickly pear cactus. Burrowing southern plains woodrats (*Neotoma micropus*) are common in these localities and have been identified as a reservoir host (Kerr et al., 1995). The parasite has been detected in sandflys (*Lutzomyia anthophora*) occupying woodrat nests (McHugh et al., 1993). The white-throated woodrat (*Neotoma albigula*) may serve as a reservoir in Arizona (Kerr et al., 1999). Infected dogs and cats have also been reported but it is not clear that they are true reservoir hosts (Craig et al., 1986).

Pathogenesis

Metacyclic promastigotes inoculated into the skin by a feeding sandfly enter macrophages by binding complement receptors after which they are engulfed by receptor-mediated endocytosis (Solbach and Laskay, 2000). The parasites multiply within the host cell, eventually destroying them and infecting new cells. The infection usually remains localized at the site of the insect bite where a granulomatous response develops and the lesion progresses from a small papule, possibly becoming pruritic, growing to 2 cm or more in diameter. As the parasite continues to multiply and infect new cells, the papule is destroyed and the epidermis becomes necrotic, thin and easily damaged. This can lead to ulceration of the lesion (Garcia and Bruckner, 1997).

Clinical Manifestations and Complications

Furner (1990) and McHugh et al. (1996) have reviewed the clinical manifestations of the cases of cutaneous leishmaniasis acquired in Texas. In general, disease presentation is similar to that observed in *L. mexicana*-infected patients from Latin America. Typically, a single painless lesion appears, often on the cheek or ear, in the form of a papule or nodule that eventually ulcerates. Lesions are usually well localized, attain a diameter of 1-3 cm, and resolve several days to months after diagnosis. Resolution may not require therapy. Complications included 1 case of unresolved disseminated disease of several years duration and 2 cases of nasal mucosal involvement, one of which progressed to perforate the septum.

Diagnostic Tests

Diagnosis depends on finding *Leishmania* parasites in aspirates, scraping, or biopsy specimens from the lesion. Amastigotes can be identified in stained smears of aspirated material, touch preparations, or histological sections. They are found in dermal macrophages and are small (3-5 μm in diameter), round or oval in shape with a small nucleus and an even smaller, rod-shaped kinetoplast (Koff and Rosen, 1994). Aseptically collected tissues can be inoculated into Schneider's *Drosophila* medium and examined microscopically for up to 4 weeks for the appearance of the flagellated promastigotes. Identification to species of both amastigotes and promastigotes is dependent on the use of DNA probes, isoenzyme analysis, or monoclonal antibodies (Garcia et al., 1997). The latter techniques are generally unavailable in a typical diagnostic laboratory but assistance with culture materials and identification can be obtained from the Division of Parasitic Diseases of the Centers for Disease Control and Prevention (CDC).

Treatment and Control Measures

A number of different treatments have been used, both individually and in combination, to successfully cure patients with localized cutaneous leishmaniasis. It is important to note however that one-fifth of all autochthonously acquired cases in Texas healed spontaneously with no treatment (McHugh et al., 1996). Nonhealing lesions may be treated with sodium stibogluconate (Pentostam), a pentavalent antimony compound available through the CDC Drug Service under an investigational New Drug

Protocol. This drug is considered a mainstay of leishmaniasis therapy but its use is accompanied by side effects, primarily arthralgias and elevated hepatic transamines (Melby et al., 1992). Amphotericin B and ketoconazole may be effective against lesions that do not respond to pentavalent antimony (Melby et al., 1992). Physical modalities, including local heat, cryotherapy, and surgical excision have also been employed with some success (Koff and Rosen, 1994). Finally, the CDC can be consulted for therapeutic advice. The most likely risk factor in the cases of cutaneous leismaniasis observed in Texas was the close proximity of patient residences to woodrat habitat (McHugh et al., 1996). Rodent control has proven effective in reducing human infection in other endemic countries. Consequently, decreasing the woodrat population in the immediate vicinity of homes may be desirable. This is accomplished by eliminating scrub around homes and destroying woodrat burrows and other likely rodent haunts such as abandoned buildings and scrap piles. The use of window screening with fine mesh screen and residual insecticides indoors, in combination with insect repellents outdoors, may also decrease the likelihood of sandfly bites.

Recent Advances and Contemporary Challenges

Thanks to the experimental use of inbred mouse models, numerous advances have been made in recent years concerning the biology of the *Leishmania* parasites, their transmission, pathogenesis, survival in vertebrate hosts, host immunity, and the genetics of resistance. With regard to *Leishmania mexicana* in the U.S.A., a number of unresolved issues have been raised by McHugh et al. (1996). For example, basic biological data such as the distribution of sand fly vectors and their seasonal activity levels are needed as are more thorough surveys of suitable reservoir hosts in the southwest. The lack of an approved delayed-type hypersensitivity skin test antigen that could be used to identify exposed individuals with subclinical infections has hindered epidemiological studies. Finally, the limited number of cases in this country, in combination with the variety of therapies employed, make it difficult to evaluate the effectiveness of any particular treatment.

AMERICAN TRYPANOSOMIASIS (CHAGAS' DISEASE)

Etiologic Agent: *Trypanosoma cruzi*

Clinical Manifestations: Acute phase symptoms are usually mild and nonspecific in patients autochthonously-infected in the U.S.A. They include persistent fever, irritability, listlessness, mild anemia, lymphocytosis, and erythema.

Complications: Fatal, acute myocarditis was reported in one instance.

Mode of Transmission: While taking a blood meal, infected "kissing" (conenose) bugs belonging to the reduviid subfamily Triatominae deposit feces containing the parasite on or near the host. The parasite invades the host by way of the vector bite wound, other skin lesions, or mucous membranes.

Diagnosis: During acute infection, the parasite can sometimes be detected in stained blood smears or in buffy coat preparations. Anti-*T. cruzi* immunoglobulin may be detectable, either by immunofluorescent assay or enzyme-linked immunosorbant assay (ELISA). Polymerase chain reaction (PCR) amplification of *T. cruzi*-specific DNA sequences can detect the parasite in seronegative patients.

Drug(s) of Choice: Benznidazole or nifurtimox are the drugs of choice but are effective only during the acute stage of infection.

Reservoir Hosts: Raccoons, opossums, armadillos, cotton rats, domestic dogs, and other mammals.

Control Measures: Eliminate brush, woodpiles, and dead trees around domestic dwellings that may serve as habitat for reservoir hosts and attract triatomine vectors. Treat kennels, animal sheds and coops with residual insecticide to eliminate vectors attracted to domestic animals. Use window screens to prevent entry of vectors and, if necessary, treat home with residual insecticides.

Etiology/Natural History

Trypanosoma cruzi, the hemoflagellate responsible for Chagas' disease, is a major cause of heart disease in Central and South America. The parasite infects 18 million people in Latin America and kills thousands of people every year (WHO, 1991). Many of the victims are children. In contrast, only 5 cases of autochthonous *T. cruzi* infection of persons in the U.S.A. have been reported (Herwaldt et al., 2000) and a recent serosurvey of blood donors from the southeast found a negligible prevalence of anti-*T. cruzi* antibody (Barrett et al., 1997). In fact, direct transmission of the parasite via transfusion of blood from asymptomatic, chronically infected Latin American immigrants represents a greater risk than autochthonous infection in this country (Leiby, et al., 1997). This is surprising because the parasite is common in triatomine bugs (Wood and Wood, 1964; Pung et al., 1995) and a variety of wild mammals (John and Hoppe, 1986) in the southern half of the U.S.A. In some localities, the prevalence of the parasite in raccoons exceeds 50% (Yabsley et al., 2001) and reports of infected dogs (Meurs et al., 1998) are not uncommon. It is not unusual for the parasite to infect nonhuman primates at animal facilities (Pung et al., 1998).

The reasons why Chagas' disease in humans is so rare in the U.S.A. are not entirely clear. It has been suggested that the lack of suitable domestic dwellings for local triatomine bugs, in combination with their zoophilicity and delayed defecation times (Zeledón, 1974), the low virulence of indigenous *T. cruzi* parasites (Kagan et al., 1966), and possible misdiagnosis of infection (Milei et al., 1992) may all play a role.

Tyler and Engman (2001) have reviewed the life cycle of *T. cruzi*. Triatomine bugs become infected when consuming mammal blood containing a flagellated form of the parasite called a trypomastigote. Trypomastigotes pass into the bug's midgut and differentiate into amastigotes that replicate and differentiate into elongate epimastigotes. Epimastigotes attach to the cuticle of the hindgut wall where they differentiate into infective metacyclic trypomastigotes. When the bug takes a blood meal from a new host it defecates and deposits the parasites on the host. The area around the bite wound is pruritic, thus inducing the host to rub the parasite-laden bug feces into the bite wound, other skin lesions, or mucous membranes. Metacyclic trypomastigotes can infect a wide range of nucleated mammalian cells including neurons, reticuloendothelial cells, and smooth, skeletal and cardiac muscle cells. Once inside a cell, the parasite differentiates into a nonflagellated amastigote and multiplies in the cytoplasm to form a pseudocyst. In the pseudocyst, amastigotes differentiate into trypomastigotes that escape into the bloodstream to infect new cells or be taken up by another triatomine bug.

Epidemiology

The transmission of *T. cruzi* to humans in Latin America is usually associated with the bite of an infected triatomine bug and the proximity of infected animal reservoirs or humans. Most of the important triatomine vectors have domestic habits and can be found in large numbers hidden in roof thatching and in the cracks of the mud-walled homes of the rural poor in Latin America. In the U.S.A., 5 cases of autochthonously acquired *T. cruzi* infection have been reported from California, Texas, and Tennessee since 1955 (Herwaldt et al., 2000). As in Latin America, all of these cases are suspected to be associated with the bite of infected triatomines because the patients had not received blood transfusions or traveled outside of the country to endemic regions. In addition, patients or their relatives sometimes complained of being bitten by triatomines, infected triatomines were sometimes found in the home, and infected raccoons and domestic dogs were sometimes found near the home (Woody and Woody, 1955; Schiffler et al., 1984; Herwaldt et al., 2000).

Pathogenesis

During the acute phase of infection, pathology is associated with the presence of the parasite. Amastigotes are easily demonstrated in diseased tissues and damage is probably due to a combination of factors including the disruptive effects of the parasite on cellular metabolism and the rupture of infected cells (Tanowitz et al., 1992). The etiology of chronic Chagas' disease is not completely understood and there is still debate over whether the disease mechanisms are autoimmune in nature or due to inflammatory responses targeting the parasite. The autoimmune hypothesis is supported by the presence of anti-self antibodies and lymphocytes in *T. cruzi*-infected hosts as well as signs of disease and intense inflammation in tissues demonstrating little or no histological evidence of the parasite (Kalil and Cunha-Neto, 1996). The alternative view is supported by the consistent detection of parasite antigens and DNA in diseased tissues examined by sensitive immuno-histochemical or PCR-based techniques and the finding that infections in chronically infected individuals are reactivated by immunosuppression (Tarleton and Zhang, 1999).

Clinical Manifestations and Complications

Acute phase symptoms are generally mild and nonspecific. They may include inflammation at the site of parasite entry, subcutaneous edema, persistent fever, irritability, listlessness, mild anemia, lymphocytosis, erythema, and enlargement of the liver, spleen, and lymph nodes. Mortality due to myocarditis and/or meningoencephalitis may occur during the acute phase, particularly in children. For example, an autochthonously infected infant from south Texas died of acute mycocarditis within days of developing symptoms (Ochs et al., 1996). More typically, in endemic areas in Latin America untreated individuals recover from the acute phase and enter an asymptomatic phase that lasts from years to decades and that may eventually develop into symptomatic Chagas' disease characterized by cardiac and gastrointestinal involvement. In these patients the progressive loss of the integrity of both the myocardium and the conduction systems results in cardiac arrhythmias, ventricular aneurysm, congestive heart failure, thromboembolism, and sudden cardiac death (Tanowitz et al., 1992; Rassi et al., 2000). In some patients, denervation of the gastrointestinal tract results in megacolon and megaesophagus (Tanowitz et al., 1992).

Diagnostic Tests

During the acute phase of infection, *T. cruzi* trypomastigotes may be detected in Giemsa-stained thin or thick blood smears or in buffy coat preparations. Trypomastigotes are approximately 20 μm in length and often assume a C or U shape on stained smears. They have a readily visible flagellum, an undulating membrane, a centrally located nucleus, and a posteriorly located kinetoplast (Garcia and Bruckner, 1997). Pseudocysts containing amastigotes may be identified in biopsy specimens. Amastigotes are small (1.5 - 4 μm), round or ovoid with a nucleus and a rod-shaped kinetoplast (Orihel and Ash, 1995). Hemoculture is sometimes used diagnostically but cultures must be examined microscopically over the course of several weeks. The parasite is rarely seen in the blood of chronically infected patients and indirect immunofluorescence or ELISA assays are commonly used to detect anti-*T. cruzi* immunoglobulin in the serum of these individuals. PCR amplification of *T. cruzi*-specific DNA sequences can detect the presence of the parasite in seronegative, and blood culture negative patients (Herwaldt et al., 2000). The Division of Parasitic Diseases of the CDC can be contacted for diagnostic assistance.

Treatment and Control Measures

Benznidazole or nifurtimox, available through the CDC Drug Service under an investigational New Drug Protocol, are the drugs of choice. They are effective only during the acute stage of infection. The efficacy of these drugs is variable and both have serious side effects including anorexia, vomiting, peripheral polyneuropathy, and allergic dermopathy (Urbina, 1999). The only therapy available for chronically infected individuals is the management of symptoms.

Triatomine bugs are attracted to locales that are inhabited by wild and domestic animals and that offer plenty of cracks and crevices where the bugs can hide during the day. These include nests and burrows of rodents, raccoons, opossums and other small mammals, dog kennels and poultry coops. The close proximity of any such habitats to human dwellings would increase the risk of people being exposed to infected bugs. Consequently, it is advisable to clear brush, dead trees, and woodpiles from around the house to discourage the presence of wild reservoir hosts, to raze abandoned buildings that may be rodent infested, and to spray the dwellings of domestic animals with residual insecticides (pyrethroids). The triatomine bugs of the U.S.A. are less likely to visit peoples' homes than their Latin American cousins but the use of window screens and residual insecticides should discourage the occasional invader. Also, while there is no evidence that humans in this county have become infected following exposure to the blood of infected reservoirs, hunters and trappers might be wise to wear latex gloves when skinning or butchering raccoons, opossums, and other known animal hosts.

Recent Advances and Contemporary Challenges

Thanks to the efforts of investigators in this country, Latin America and other parts of the world, our understanding of host immunity to *T. cruzi* and the pathogenic mechanisms associated with Chagas' disease has improved dramatically. However, better treatments for the disease and an effective vaccine are desperately needed.

BALANTIDIASIS

Etiologic Agent: *Balantidium coli*

Clinical Manifestations: Diarrhea, abdominal pain, and colitis are often apparent though many infected patients are asymptomatic.
Complications: Ulceration of the colon, dysentery.
Mode of Transmission: People become infected by swallowing cysts in feces from infected reservoirs or humans. Most commonly, the cysts are in contaminated food and water.
Diagnosis: Demonstration of trophozoites, and occasionally cysts, in stool specimens; demonstration of trophozoites in scrapings or biopsy specimens of colonic ulcers.
Drug(s) of Choice: Tetracycline, iodoquinol or metronidazole (= flagyl)
Reservoir Hosts: Pigs, rodents, cats, and nonhuman primates.
Control Measures: Good sanitation and personal hygiene, prevention of human-pig contact, prevention of environmental contamination by pig feces.

Etiology/Natural History

Balantidium coli has the distinction of being the largest and only ciliated protozoan parasite of humans. The parasite is observed as a harmless commensal in hogs though it is also reported from a variety of other wild and domestic mammals. Most investigators consider balantidiasis to be a true zoonotic disease and regard *Balantidium* parasites from pigs and humans to be the same species. However, some pig strains of the parasite can be immunologically distinguished from human strains and are not infectious to humans (Zaman, 1998).

Cysts are the infective form of the parasite and are transmitted to new hosts by the fecal-oral route. The trophozoite stage of the parasite excysts in the small intestine and multiplies in the lumen and the wall of the colon. Cysts are only observed in fecal material, not tissue specimens. Both trophozoites and infective cysts are passed in the feces. Cysts are seen in normal fecal specimens while trophozoites are most likely to be observed in patients with diarrhea or dysentery (Orihel and Ash, 1995). Trophozoites do not survive outside the body but the environmentally resistant cysts contaminate food and water to be passed on to new hosts (Zaman, 1998).

Epidemiology

Balantidium coli is common in pigs throughout the world, including those raised in the U.S.A. (Morris et al., 1984). Humans are apparently not susceptible to some pig strains of the parasite though human infection is most common where there is a close association between people and pigs (Orihel and Ash, 1995). Balantidiasis is most frequently reported from tropical and subtropical regions. The disease is only rarely observed in the U.S.A. though it has been reported in psychiatric units in this country where transmission is probably due to human-to-human contact, poor hygiene, and coprophagy (Zaman, 1998).

Pathogenesis

The pathogenic mechanisms associated with severe, acute balantidiasis are not clear but are associated with invasion and inflammation of the colon. Susceptibility to infection, immune and nutritional status, and age may be important contributing factors. The parasite secretes hyaluronidase which is likely to facilitate its invasion of the mucosa and submucosa with subsequent formation of ulcers of varying degrees of severity, abscesses, and hemorrhagic lesions (Tempelis and Lyseko, 1957; Castro et al., 1983).

Clinical Manifestations and Complications

Most infected people are asymptomatic carriers who pass infective cysts in their feces, thus transmitting the parasite to others where personal hygiene is poor. The most common disease symptoms include chronic recurrent diarrhea alternating with constipation (Zaman, 1998). Acute balantidiasis is a rare but potentially fatal condition characterized by severe dysentery with profound prostration, fever, vomiting, dehydration, and shock. Complications include perforation of the bowel and extracolonic involvement including infection of the mesenteric lymph nodes, peritoneum, lungs, liver, and appendix (Castro et al., 1983; Dodd, 1991).

Diagnostic Tests

Balantidium trophozoites stain poorly and are most easily recognized in wet mounts of fresh or concentrated feces (Garcia and Bruckner, 1997). Trophozoites are distinguished from other intestinal protozoa by their large size (50 - 200 μm in length by 40 - 70 μm in width), and the presence of cilia (Orihel and Ash, 1995). The cysts are occasionally encountered in feces but not in tissue specimens. They are spherical, 50 - 70 μm in diameter, and display a prominent kidney-bean-shaped macronucleus when stained (Orihel and Ash, 1995; Zaman, 1998). Depending on the plane of section, trophozoites in histological sections may display cilia, the macro- and the micronucleus, a feeding cytostome, and contractile vacuoles (Orihel and Ash, 1995).

Treatment and Control Measures

The drug of choice for balantidiasis is tetracycline though it is contraindicated in pregnancy and in children younger than 8 years (The Medical Letter, 2000). Iodoquinol and metronidazole are alternative treatments. All 3 drugs are approved by the Food and Drug Administration but are considered investigational for the treatment of this disease. Since *Balantidium* is transmitted by the ingestion of fecal cysts, good personal hygiene is an important preventative measure. Hygienic rearing and slaughtering of pigs to prevent contamination of the environment by pig feces is equally important. Treatment of infected humans will eliminate person-to-person transmission (Zaman, 1998).

Recent Advances and Contemporary Challenges

A number of interesting questions concerning *Balantidium* infection are still unanswered. For example, it is not known if the different clinical manifestations of the disease, which range from asymptomatic to severe dysentery with tissue invasion, are the result of differences in host susceptibility or the virulence of particular strains of the parasite. Consequently, biological and molecular characterization of strains of *Balantidium* from reservoir hosts and from humans patients are warranted as are genetic and immunological studies concerning variation in susceptibility of infected humans to the parasite.

GIARDIASIS

Etiologic Agent: *Giardia lamblia* (also known as *Giardia intestinalis* or *Giardia duodenalis*).

Clinical Manifestations: Asymptomatic cyst passage; acute, self-limited diarrhea; chronic diarrhea, malabsorption, and weight loss. The most common symptoms include: diarrhea, malaise, flatulence, foul-smelling stools, abdominal cramps, bloating, nausea, anorexia, weight loss, fever, vomiting.

Complications: Chronic diarrhea, malabsorption and weight loss, growth impairment in children

Mode of Transmission: Drinking contaminated water or food, human to human spread, animal to human transmission

Diagnosis: Identification of trophozoites or cysts on direct stool smear examination, *G. lamblia* antigen detection in stool specimens or duodenal fluid using immunofluorescent antibody or EIA, detection of anti-*Giardia* antibody in serum

Drug(s) of Choice: Metronidazole is the drug of choice. Tinidazole, furazolidone, albendazole, and paromomycin are alternative choices.

Reservoir Hosts: Humans are the principal reservoir hosts, Dogs, cats, beavers, gerbils, cattle, and sheep also may be infected.

Control Measures: Adequate filtration, chlorination, and maintenance of water distribution systems. Improved sanitation and personal hygiene in adult and child care centers. Travelers to developing countries or the wilderness should boil drinking water.

Etiology/Natural History

Giardia has a worldwide distribution and is the most commonly identified intestinal parasite in man. They are flagellate protozoa that exist in the trophozoite or freely living stage and the cyst or infectious stage. Several species of *Giardia* exist that infect a wide variety of mammals, although humans are the principal reservoirs of infection. *Giardia lamblia* (also called *G. intestinalis* or *G. duodenalis*) infect humans. Infection occurs following ingestion of cysts that survive the gastric acid of the stomach. Excystation occurs in the proximal small intestine and the emerging trophozoites attach to the epithelium of the proximal small intestine. Trophozoites multiply and subsequently encyst in the distal small intestine and large intestine. As few as 10 to 25 cysts may establish infection while 150 to 20,000 cysts may be excreted per gram of stool daily. Cysts may remain viable in water for as long as two months.

Epidemiology

Giardia has a worldwide distribution with prevalence rates from 2 to 4 percent in developed countries to 20 to 30 percent in developing countries. Rates as high as 20 percent have been found in certain areas of the United States. People most commonly become infected by ingestion of contaminated water or food but also by hand to mouth transfer of cysts from the feces of an infected person. Those at highest risk of infection include infants and young children, especially those in day care centers, sexually active male homosexuals, persons in custodial institutions, and immunocompromised persons. *Giardia* cysts survive well in the environment, especially in cold water. Waterborne outbreaks are common in mountainous regions such as ski resorts and in wilderness areas or among travelers to third world countries where water purification may be substandard. There is compelling evidence that giardiasis is a zoonosis. *Giardia* isolates from humans and animals have been shown to be genotypically indistinguishable. Epidemiologic studies have strongly suggested that both wild and domestic animals serve as an infectious reservoir for humans (Issac-Renton, et al., 1993; Craun, 1984, Warburton et al., 1994).

Pathogenesis

The pathogenic mechanisms responsible for diarrhea and malabsorption associated with *Giardia* infection are not entirely clear but appear to be related to the number and virulence of the *Giardia* strain type ingested and the host's immune response to infection. Structural damage of the small intestinal brush border with concomitant diasaccharidase deficiency and inflammatory infiltration have been well documented (Katelaris and Farthing, 1992; Balazs and Szaltocky, 1978; Welsh et al., 1984). Both humoral and lymphocyte and macrophage immune responses to infection have been demonstrated (denHollander et al., 1988). Mucosal invasion appears to be rare and no evidence exists for an enterotoxin.

Clinical Manifestations and Complications

The incubation period of *Giardia* from ingestion of cysts until the development of symptoms is usually 1 to 2 weeks. A broad spectrum of clinical manifestations occur including asymptomatic passage of cysts, acute self-limited diarrhea, and a chronic syndrome of diarrhea, malabsorption, and weight loss. Only about 25 to 50 percent of those infected will become symptomatic but the majority of these people will experience a diarrehal syndrome lasting one to several weeks. Typical symptoms include diarrhea, abdominal cramps bloating, malaise, flatulence, nausea, anorexia, and weight loss. Fever and vomiting are less common. Stools may initially be watery but later are greasy and foul smelling. Blood and pus usually are absent. Prolonged diarrhea of greater than a week and weight loss are characteristic and help to distinguish giardiasis from bacterial or viral gastroenteritis (Moore et al., 1969). Cyst passage may last as long as 6 months (Pickering et al., 1984).

Diagnostic Tests

Giardiasis should be considered in all patients experiencing prolonged diarrhea, especially when accompanied by malabsorption and weight loss. A history of immunosuppression, travel to a third world country or a wilderness area, or contact with children in a day care center or adults in custodial care should further raise suspicion of giardial infection. Giardia cysts and trophozoites can be detected by light microscopy in stool or duodenal specimens. Use of an immunofluorescent antibody (IFA) stain against cyst proteins or an enzyme immunoassay (EIA) to detect *Giardia* antigens increases sensitivity. Appropriate stool examination will establish the diagnosis in 50 to 70 percent of patients by a single examination, 85 percent by a second specimen and greater than 90 percent by a third (Niak et al., 1978). Serologic testing is useful in research studies and seldom is used for clinical diagnosis.

Treatment and Control Measures

Patients experiencing acute or chronic symptoms of giardiasis should be treated. Asymptomatic cyst passers generally are not treated except to prevent transmission in households of patients with hypogammaglobulinemia or cystic fibrosis and to prevent transmission from toddlers to pregnant women. Because antibiotic susceptibility testing is difficult treatment recommendations have been based on clinical experience. Metronidazole (Flagyl) is the drug of choice. It is given for a 5 to 7 day course and has an 80 to 95 percent cure rate. Tinidazole, furazolidone, albendazole, and paromomycin are alternative choices. If therapy fails, a course can be repeated with the same drug. Relapse is common in immunocompromised patients and treatment may need to be prolonged or combination therapy used. Prevention requires proper filtration, chlorination, and maintenance of water distribution systems. Adequate sanitation and personal hygiene should be maintained in adult and child care centers. Travelers to developing countries or the wilderness should boil drinking water.

MICROSPORIDIOSIS

Etiologic Agents: *Encephalitozoon* spp., *Enterocytozoon bieneusi,* and others

Clinical Manifestations: Infection is generally asymptotic in immunocompetent individuals. Disease manifestations occur most frequently in AIDS patients and include chronic diarrhea with wasting, fever, malaise, keratoconjunctivitis, biliary disease, and infection of the respiratory and genitourinary tracts.

Mode of Transmission: The modes of transmission are uncertain. Consumption of spores in contaminated food or water, inhalation of spores, autoinoculation of the eye, person-to-person contact, and sexual transmission are suspected.

Diagnosis: Light microscopic detection of spores in stained histological preparations, stools, and urine. Electron microscopy is required for identification to species level.

Drug(s) of Choice: Albendazole is the drug of choice for intestinal and disseminated microsporidiosis though it is not effective against *E. bieneusi.* Oral albendazole plus fumagillan eye drops are recommended for ocular microsporidiosis.

Reservoir Hosts: Wild and domestic mammals, birds and other animals possibly serve as reservoir hosts.

Control Measures: Environmental sources of the parasite and its modes of transmission to humans are not certain. Consequently, effective preventative measures are difficult to devise. Good personal hygiene, particularly hand washing following contact with human or animal urine, feces, or respiratory secretions may be beneficial.

Etiology/Natural History

The microsporidia are a diverse group of over 1000 species of intracellular spore-forming parasites. Though long known to infect many invertebrates, insects in particular, and all classes of vertebrates, microsporidia were rarely reported to cause disease in humans until the advent of the AIDS pandemic (Wittner, 1999). Microsporidia are now recognized as important opportunistic pathogens in immunocompromised individuals. *Enterocytozoon bieneusi* is the most commonly reported microspordium in humans and up to 50% of AIDS patients with chronic diarrhea may be infected with this parasite (Bryan, 1995). Other important species include *Encephalitozoon hellum*, *Encephalitozoon intestinalis* and *Encephalitozoon cuniculi* (Mathis 2000).

The microsporidia have a number of striking features that are unusual in a eukaryotic organism including an extremely small genome that lacks introns, no mitochondria, no centrioles, and prokaryote-like ribosomes (Weiss, 2001). Due to these traits, microsporidia have been traditionally considered primitive protozoa that evolved prior to the endosymbiotic acquisition of mitochondria. However, recent studies suggest that they are actually highly derived fungi, or relatives of the fungi, that underwent genetic and functional losses as they evolved (Mathis, 2000, Weiss, 2001).

Another unique feature of microsporidia is a polar filament coiled around the interior of the spore. The filament can be explosively discharged to pierce a host cell and inject the sporoplasm of the parasite into the host cell cytoplasm. After entering the host cell the microsporidia begin a proliferative phase and then differentiate into spores. The spores are released and either invade new cells or are disseminated into the environment (Didier et al., 1998; Weiss, 2001).

Epidemiology

The sources of microsporidia spores in the environment are assumed to be feces, urine or respiratory secretions from infected people or animals. People may be exposed to spores of either zoonotic or human origin by the fecal-oral or urinary-oral routes, inhalation of spores in respiratory secretions or urine, or by autoinoculation of the eye (Bryan, 1995). Definitive proof that these parasites are zoonotic is lacking but indirect evidence suggests that some of the microsporidia might occasionally be transmitted to humans from animal reservoir hosts. For example, natural infections of *E. cuniculi*, *E. intestinalis*, *E. hellem* and *E. bieneusi* have been found in wild and domestic animals. Some of these animal strains of the parasite are genetically similar or identical to parasites found in humans suggesting a potential lack of species barriers (Mathis et al., 2000; Rinder et al., 2000). Risk factors associated with intestinal microsporidiosis include exposure to livestock, fowl, and pets (Weiss, 2000), male homosexuality, and close contact among HIV-infected persons (Mathis, 2000).

Pathogenesis

The pathogenesis of microsporidiosis is not well understood. Gastrointestinal infections result in mucosal damage ranging in severity from normal histological appearance to gross villous atrophy resulting in nutrient malabsorption and osmotic diarrhea (Conteas et al., 2000). Enterocyte damage may be directly due to parasite invasion and rupture of host cells (Goodgame, 1996). However, fecal tumor necrosis factor-α is elevated in microsporidia-infected AIDS patients and may contribute to pathogenesis

(Didier et al., 1998). The pathogenesis of microsporidia-related cholangitis is not known but may be due to desquammation of bile duct epithelium exposing the lamina propria to bile salts and other compounds and inducing inflammation and fibrosis (Kotler and Orenstein, 1998).

Clinical Manifestations and Complications

The clinical manifestations of microsporidiosis vary depending on the infecting species but are almost always associated with HIV infection (Bryan, 1995; Schwartz et al., 1996). *Enterocytozoon bieneusi* infections are typically characterized by chronic diarrhea, dehydration and weight loss. Biliary infection occurs in some patients and may result in cholecystitis with abdominal pain, fever, nausea and vomiting. Similar symptoms are associated with *E. intestinalis*. Persons infected with *E. hellum* and *E. cuniculi* suffer keratoconjunctivitis. Other diseases associated with *Encephalitozoon* species include bronchiolitis, sinusitis, nephritis, cystitis, and ureteritis (Bryan, 1995).

Diagnostic Tests

Light microscopic detection of spores in stained clinical specimens is the standard method of diagnosing microsporidiosis. The spores are extremely small (1-5 µm), oval, pyriform, or elongate (Orihel and Ash, 1995). Several staining methods have been employed. A modified trichrome stain containing 10 times the normal concentration of chromotrope 2R is a widely used method for staining spores in stool specimens. This technique colors the spore wall bright pinkish red. A pinkish red stripe may be visible in the middle of the spore. Spores are acid fast variable and have a small periodic acid-Schiff positive posterior body (Garcia and Bruckner, 1997). Since the spores are so small, species-specific morphological features such as thickness of the spore wall and the number of polar filament coils cannot be readily visualized with the light microscope. Consequently, electron microscopy is the "gold standard" for confirming infection and for identification of the parasites to the species level (Orihel and Ash, 1995). Immunofluorescent and PCR-based detection methods are now being developed (Müller et al, 1999).

Treatment and Control Measures

The treatment of microsporidiosis is challenging because a practically impervious proteinaceous exospore and a chitinous endospore protect the extracellular spore form of the parasite. Furthermore, the intracellular form of the parasite is dependent on the host cell for much of its energy and reproductive requirements. Albendazole, a benzimidazole derivative with anthelmintic and antifungal activity, is the drug of choice for intestinal and disseminated microsporidiosis (The Medical Letter, 2000). Unfortunately, albendazole is not very effective against *E. bieneusi*. Preliminary studies suggest that oral fumagillin and its analogue, TNP-470, may be good alternative therapies for this particular parasite (Conteas et al., 2000). Oral albendazole plus fumagillan eye drops are recommended for ocular microsporidiosis. Fumagillan drops are prepared from Fumidil-B, a product used to treat microsporidiosis in honeybees (The Medical Letter, 2000). Interestingly, the prevalence of microsporidiosis in AIDS patients has declined dramatically over the past several years, possibly due to enhancement of immunity following highly effective antiretroviral therapy using protease inhibitors (Conteas et al., 2000).

The sources and the transmission modes of human microsporidiosis are uncertain. Thus, it is difficult to predict the most effective means of preventing infection. However, since spores are found in body fluids, good personal hygiene, particularly hand washing following contact with human or animal urine, feces, or respiratory secretions may be beneficial.

RECENT ADVANCES AND CONTEMPORARY CHALLENGES

Though recent studies have advanced our knowledge of the basic biology, genetics, and systematics of the microsporidia, much of the natural history of these parasites and the pathogenesis of microsporidiosis remain unclear. Until epidemiological studies are performed to determine the sources of infection, modes of transmission, and risk factors associated with microsporidiosis, the development of effective control and prevention measures will be difficult to implement.

REFERENCES

Barrett VJ, Leiby DA, Odom JL, et al. 1997. Negligible prevalence of antibodies against *Trypanosoma cruzi* among blood donors in the southeastern United States. *Am J Clin Pathol* 108: 499-503.

Bryan RT. 1995. Microsporidiosis as an AIDS-related opportunistic infection. *Clin Infect Dis* 21: S62-65.

Despommier DD, Gwadz RW, Hotez PJ, Knirsch CA. 2000. *Parasitic Diseases*. New York, NY: Apple Trees Productions, LLC. 345 pp.

Didier ES, Snowden KF, Shadduck JA. 1998. Biology of microsporidian species infecting mammals. *Adv Parasitol* 40: 283-320.

Dodd LG. 1991. *Balantidium coli* as a cause of acute appendicitis. *J Infect Dis* 163: 1392.

Castro J, Vazquea-Iglesias JL, Arnal-Monreal F. 1983. Dysentery caused by *Balantidium coli* - report of two cases. *Endoscopy* 15: 272-274.

Conteas CN, Berlin OGW, Ash LR, Pruthi JS. 2000. Therapy for human gastrointestinal microsporidiosis. *Am J Trop Med Hyg* 63: 121-127.

Craig TM, Barton CL, Mercer SH, et al. 1986. Dermal leishmaniasis in a Texas cat. *Am J trop Med Hyg* 35: 1100-1102.

Enserink M. 2000. Has leishmaniasis become endemic in the U.S.? *Science* 290: 1881-3.

Furner BB. 1990. Cutaneous leishmaniasis in Texas: Report of a case and review of the literature. *J Amer Acad Dermatol* 23: 368-371.

Garcia LS, Bruckner DA. 1997. *Diagnostic Medical Parasitology*. Washington, DC: ASM Press.
Goodgame RW. 1996. Understanding intestinal spore-forming protozoa: cryptosporidia, microsporidia, isospora, and cyclospora. *Ann Intern Med* 124: 429-441.
Herwaldt BL, Grijalva MJ, Newsome AL, et al. 2000. Use of polymerase chain reaction to diagnose the fifth reported US case of autochthonous transmission of *Trypanosoma cruzi*, in Tennessee, 1998. *J Infect Dis* 181: 395-399.
John DT, Hoppe KL. 1986. *Trypanosoma cruzi* from wild raccoons in Oklahoma. *Am J Vet Res* 47: 1056-9.
Kagan IG, Norman L, Allain D. 1966. Studies on *Trypanosoma cruzi* isolated in the United States: A review. *Rev Biol Trop* 14: 55-73.
Kalil J, Cunha-Neto E. 1996. Autoimmunity in Chagas disease cardiomyopathy: fulfilling the criteria at last? Parasitol Today 12: 396-399.
Kerr SF, McHugh CP, Dronen NO. 1995. Leishmaniasis in Texas: Prevalence and seasonal transmission of *Leishmania mexicana* in *Neotoma micropus*. *Am J Trop Med Hyg* 53: 73-77.
Kerr SF, McHugh CP, Merkelz R. 1999. Short report: A focus of *Leishmania mexicana* near Tucson, Arizona. *Am J Trop Med Hyg* 61: 378-379.
Koff AB, Rosen T. 1994. Treatment of cutaneous leishmaniasis. *J Amer Acad Dermatol* 31: 693-708.
Kotler DP, Orenstein JM. 1998. Clinical syndromes associated with microsporidiosis. *Adv Parasitol* 40: 321-349.
Leiby DL, Read EJ, Lenes BA, et al. 1997. Seroepidemiology of *Trypanosoma cruzi*, etiologic agent of Chagas' disease in US blood donors. *J Infect Dis* 176: 1047-1052.
Mathis A. 2000. Microsporidia: emerging advances in understanding the basic biology of these unique organisms. *Int J Parasitol* 30: 795-804.
The Medical Letter. 2000. *Drugs for Parasitic Infections*. New Rochelle, NY: The Medical Letter, Inc.
Melby PC, Kreutzer RD, McMahon-Pratt D, et al. 1992. Cutaneous leishmaniasis: review of 59 cases seen at the National Institutes of Health. *Clin Infect Dis* 15: 924-37.
Meurs KM, Anthony MA, Slater M, Miller MW. 1998. Chronic *Trypanosoma cruzi* infection in dogs: 11 cases (1987-1996). *J Am Vet Med Assoc* 213: 497-500.
McHugh CP, Grogl M, Kreutzer RD. 1993. Isolation of *Leishmania mexicana* (Kinetoplastida: Trypanosomatidae) from *Lutzomyia anthophora* (Diptera: Psychodidae) collected in Texas. *J Med Entomol* 30: 631-633.
McHugh CP, Melby PC, LaFon SG. 1996. Leishmaniasis in Texas: epidemiology and clinical aspects of human cases. *Am J Trop Med Hyg* 55: 547-55.
Milei J, Mautner B, Storino R, et al. 1992. Does Chagas' disease exist as an undiagnosed form of cardiomyopathy in the United States? *Am Heart J* 123: 1732-5.
Morris RG, Jordan HE, Luce WG, et al. 1984. Prevalence of gastrointestinal parasitism in Oklahoma swine. *Am J Vet Res* 45: 2421-23.
Müller A, Stellerman K, Hartman P, et al. 1999. A powerful DNA extraction method and PCR for detection of microsporidia in clinical stool specimens. *Clin Diag Lab Immunol* 6: 243-6.
Ochs DE, Hnilica VS, Moser DR, et al. 1996. Postmortem diagnosis of autochthonous acute Chagasic myocarditis by polymerase chain reaction amplification of a species-specific DNA sequence of *Trypanosoma cruzi*. *Am J Trop Med Hyg* 54: 526-9.
Orihel TC, Ash LR. 1995. *Parasites in Human Tissues*. Chicago, IL: ASCP Press, 1995.
Pung OJ, Banks CW, Jones DN, Krissinger MW. 1995. *Trypanosoma cruzi* in wild raccoons, opossums, and triatomine bugs in southeast Georgia, U.S.A. *J Parasitol* 81: 324-326.
Pung OJ, Spratt J, Clark CG, et al. 1998. *Trypanosoma cruzi* infection of free-ranging lion-tailed macaques (*Macaca silenus*) and ring-tailed lemurs (*Lemur catta*) on St. Catherine's Island, Georgia, USA. *J Zoo Wildlife Med* 29: 25-30.
Rassi A. Jr, Rassi A, Little, WC. 2000. Chagas' heart disease. Clin Cardiol 23: 883-889.

Rinder H, Thomschke A, Dengjel B, et al. 2000. Close genetic relationship between *Enterocytozoon bieneusi* from humans and pigs and first detection in cattle. *J Parasitol* 86: 185-188.

Schiffler RJ, Mansur GP, Navin TR, Limpakarnjanarat K. 1984. Indigenous Chagas' disease (American trypanosomiasis) in California. *J Amer Med Assoc* 251: 2983-2985.

Roberts LS, Janovy J, Jr. 1996. *Gerald D. Schmidt and Larry S. Roberts' Foundations of Parasitology, 5th ed.* Dubuque, IA: Wm. C. Brown Pub.

Schwartz DA, Sobottka I, Leitch GJ, et al. 1996. Pathology of microsporidiosis: emerging parasitic infections in patients with acquired immune deficiency syndrome. *Arch Pathol Lab Med* 120: 173-187.

Solbach W, Laskay T. 2000. The host response to *Leishmania* infection. *Adv Immunol* 74: 275-317.

Tanowitz HB, Kirchhoff LV, Simon D, et al. 1992. Chagas' disease. *Clin Microbiol Rev* 5: 400-419.

Tarleton RL, Zhang L. 1999. Chagas disease etiology: autoimmunity of parasite persistence? *Parasitol Today* 15: 94-99.

Tempelis CH, Lysenko MG. 1957. The production of hyaluronidase by *Balantidium coli*. *Exp Parasitol* 6: 31-37.

Tyler KM, Engman DM. 2001. The life cycle of *Trypanosoma cruzi* revisited. *Int J Parasitol* 31: 472-481.

Urbina JA. 1999. Chemotherapy of Chagas' disease: the how and the why. *J Mol Med* 77: 332-338.

Weiss LM. 2001. Microsporidia: emerging pathogenic protists. Acta Tropica 78: 89-102.

WHO. 1991. Control of Chagas' disease. *WHO Tech Rep Series* 811: 1-95.

Wittner M. 1999. Historic perspective on the microsporidia: expanding horizons. In: Wittner M (Ed.), *The Microsporidia and Microsporidiosis.* Washington D. C.: ASM Press.

Wood SF, Wood FD. 1964. New locations for Chagas' trypanosome in California. *Bull So Calif Acad Sci* 63: 104-111.

Woody NC, Woody HB. 1955. American trypanosomiasis (Chagas' disease): first indigenous case in the United States. *J Amer Med Assoc* 159: 676-677.

Yabsley MJ, Noblet GP, Pung OJ. 2001. Comparison of serological methods and blood culture for the detection of *Trypanosoma cruzi* infection in raccoons (*Procyon lotor*). *J Parasitol* 87: 1155-1159.

Zaman V. 1998. *Balantidium coli.* In: Topley WWC (Ed.), *Topley and Wilson's Microbiology and Microbial Infections*, 9th Edition. New York, NY: Oxford University Press.

Zeledón R. 1974. Epidemiology, modes of transmission and reservoir hosts of Chagas' disease. In: *Trypanosomiasis and Leishmaniasis with Special Reference to Chagas' Disease*, Ciba Fnd Symp 20. Amsterdam, Netherlands: Associated Scientific Publishers.

ZOONOTIC ARTHROPOD PARASITES

Lance A. Durden

**Institute of Arthropodology and Parasitology,
Georgia Southern University, Statesboro, Georgia**

BEDBUGS, BATBUGS AND BIRDBUGS (INSECTA: CIMICIDAE)

Etiologic Agents: Mainly *Cimex lectularius* (bedbug) (Figure 1) but also *Cimex pilosellus* (batbug), *Haematosiphon inodorus* (Mexican chicken bug), *Oeciacus vicarius* (swallow bug) and *Cimexopsis nyctalis* (a bug associated with chimney swifts).
Clinical Manifestations: Bites on skin; may be similar to mosquito bites.
Complications: Bite sensitivity; localized dermatologic or (rarely) generalized systemic reactions.
Mode of Transmission: From infested hosts in vicinity; typically, these are bats, chickens or other birds.
Diagnosis: Bites and location and identification of the bugs.
Drug(s) of Choice: Oral or topical antihistamine is usually effective in uncomplicated cases.
Reservoir Hosts: Several species of bats and birds, especially those that roost in or close to buildings.
Control Measures: Elimination or re-location of the reservoir hosts, treatment of infested dwellings with insecticides, and application of a topical insect repellent on the skin such as DEET (N-N-diethyl-m-toluamide).

Etiology/Natural History

Worldwide, 91 species of cimicid bugs have been described but only a handful of these occur in North America (Usinger, 1966). Cimicids feed on the blood of mammals or birds, and some will readily feed on humans. They are relatively small (5-7 mm long as adults), dorso-ventrally flattened insects with a simple (hemimetabolous) life-cycle. Nymphs resemble miniature adults except for their lack of genitalia and a few other structures. The first-stage nymph takes one or more blood meals and then molts to the next (larger) nymphal stage, and so on, for a total of about 5 nymphal stages before the molt to adulthood occurs. Adults have a unique method of mating known as "traumatic insemination" in which the male pierces the female abdominal integument at almost any site with his sharp intromittant organ. Cimicids typically hide in cracks and crevices close to the host when they are not feeding. In human dwellings, bedbugs may conceal themselves in wall and floor cracks and in furniture such as couches.

Epidemiology

The bedbug, which is distributed worldwide, typically feeds on humans or bats but will also feed on birds. This represents a zoonotic risk, especially in buildings that are occupied by roosting bats. Some elderly people, persons living alone, or those with failing eyesight, may have huge bedbug infestations in their dwellings. Bedbugs may occur anywhere but are re-emerging in some inner city areas in the presence or absence of bat reservoirs. Bedbugs usually feed when people are sleeping so patients may be unaware of infestations in their houses. The batbug will also feed on humans but it must have a source of bat hosts for long-term survival. The Mexican Chicken bug mainly feeds on poultry in Central America as its name implies, but it will also feed on humans and may be encountered in parts of the southwestern United States. The swallow bug is typically confined to the mud nests of swallows but will feed on people who disturb infested nests or rest close to them. *Cimexopsis nyctalis* occasionally feeds on people who have chimney swifts nesting in their houses.

Pathogenesis

As with most biting arthropods, reaction to a bug bites in humans, is mainly an immune response to the salivary enzymes and other biochemicals (e.g., anticoagulants, analgesics, etc.) inoculated by the bugs during blood feeding, which act as antigens.

Clinical Manifestations and Complications

The most common reaction to cimicid bites is localizied skin reactions similar to mosquito bites, but large fluid-filled bullae develop at bite sites in some individuals. Erythema occasionally occurs, mainly in response to multiple bites. Rarer reactions include severe immediate or delayed sensitivity, including anaphylaxis. Large bedbug populations may cause anemia or iron deficiency in some individuals. Irritability (from sleep deprivation) and neuroses can also occur. Some evidence suggests that bedbugs may be vectors of hepatitis B virus (Blow et al., 2001); claims of bedbugs being vectors of HIV viruses are unsubstantiated.

Diagnostic Tests

Diagnosis of skin bites or other clinical complications is usually inconclusive because bites from many different arthropods may cause similar dermatologic or systemic reactions. Therefore, clinical diagnosis often must include a careful epidemiological analysis of the patient's home to search for cimicids or their reservoir hosts.

Treatment and Control Measures

Uncomplicated bites can usually be alleviated by applying ice and the administration of oral or topical antihistamine. Clinical complications should be treated as dictated by the symptoms. Repellents such as DEET on skin, or permethrin on clothing, will prevent many bites. More efficacious and long-term control measures include treating infested premises for bug control, screening buildings against bat and bird colonization, and not sleeping or resting close to known colonies of bats, chickens or nesting birds.

Recent Advances and Contemporary Challenges

Because cimicid bites and infestations are not currently a common occurrence in North America, misdiagnosis is frequent. Chronic cases may be misdiagnosed as allergic dermatitis or other skin disorders. Because bedbug infestations appear to be re-emerging, especially in some inner cities, more accurate diagnostic tools are required. Specific antigen blood tests for detection of cimicid bites are in development. Hopla (1982) and Schofield and Dolling (1993) provide additional information on cimicids that bite humans.

CONENOSE BUGS AND KISSING BUGS (INSECTA: TRIATOMINAE)

Etiologic agents: *Triatoma* spp. (Figure 2), especially *T. sanguisuga*
Clinical Manifestations: Skin inflammation due to bites, especially if these are near the mouth or eyes.
Complications: Bite sensitivity; transmission of pathogens, most notably *Trypanosoma cruzi,* the etiologic agent of Chagas disease.
Mode of Transmission: From infested vertebrate hosts and their nests in the vicinity.
Diagnosis: Bites and location and identification of the bugs.
Drug(s) of Choice: Oral or topical antihistamine in uncomplicated cases.
Reservoir Hosts: In North America, these are mainly opossums, raccoons, woodrats and chickens and their nests or roosting sites.
Control Measures: Elimination or re-location of the reservoir hosts, insecticidal treatment of infested dwellings, and application of insect repellent on the skin.

Etiology/Natural History

Worldwide, 106 species of blood-feeding triatomine bugs have been described (Schofield, 1994). Many of these inhabit the neotropics but 5 species occur in North America especially in the southwestern United States (Ryckman and Casdin, 1976); one of these, *T. sanguisuga*, also occurs in eastern North America. These large bugs (5-45 mm long as adults) feed on various vertebrate hosts such as reptiles, birds, and mammals including humans. Like bedbugs, triatomines have a hemimetabolous life cycle with nymphal stages similar to adults but smaller. However, unlike cimicids, adult triatomines are winged and most are efficient flyers. Also, triatomines are not dorsoventrally flattened and do not use traumatic insemination during mating. Like cimicids, triatomines are typically associated with their hosts only during feeding; they conceal themselves in the host dwelling at other times. Because of their large size, triatomines can imbibe large bloodmeals during a single feeding and can survive for weeks or months between bloodmeals.

Epidemiology

Bites by triatomine bugs are much more common in Central and South America where these insects are almost ubiquitous and where they commonly transmit the agent of Chagas disease. However, North American triatomines will bite humans and, although some of them may be infected with *T. cruzi*, autochtonous Chagas disease is curiously almost non-existent north of Mexico. The bugs usually bite at night when their hosts are sleeping. In humans, they often feed around the mouth (hence one of their common names) or eyes. In partly thatched dwellings in the neotropics, bugs conceal themselves in wall cracks and in roof material after a bloodmeal.

Pathogenesis

Triatomine bites can occasionally be painful, presumably because these insects are so large. Usually no major pathology is associated with these bites.

Clinical Manifestations and Complications

Most triatomine bites present as mildly reddened areas. Sometimes the bites itch and accompanying edema and erythema may occur in mildly hypersensitive reactions. Anaphylaxis has been recorded but is rare. A major complication is the potential for transmission of *T. cruzi*. Triatomines have also been suggested as vectors of hepatitis B and HIV viruses but these claims are unsubstantiated.

Diagnostic Tests

Bites in conjunction with the location and identification of the bugs is usually diagnostic, especially if other vertebrate blood sources live in the immediate vicinity.

Treatment and Control Measures

Insecticidal applications in and around houses and animal facilities, the elimination or relocation of reservoir hosts in the vicinity, and sealing cracks in houses all help to reduce the ability of kissing bugs to locate humans. Topical insect repellents are also helpful.

Recent Advances and Contemporary Challenges

Immunotherapy to ameliorate bite reactions in humans using extracts from kissing bug salivary glands has been successfully used against some *Triatoma* species. Hopla (1982), Schofield and Dolling (1993) and Schofield (1994) provide further information on triatomine bugs.

FLEAS (INSECTA: SIPHONAPTERA)

Etiologic Agents: Mainly *Ctenocephalides felis* (cat flea) (Figure 3); also *Ctenocephalides canis* (dog flea), *Nosopsyllus fasciatus* (Northern rat flea), *Orchopeas howardi* (squirrel flea), *Xenopsylla cheopis* (Oriental rat flea), and others.
Clinical Manifestations: Small bites, often itchy, especially around the ankles.
Complications: Flea-bite dermatitis, other allergic reactions, and pathogen transmission.
Mode of Transmission: Mainly from pets or wild mammals.
Diagnosis: Bites and the presence of fleas in dwellings or on companion animals.
Drug(s) of Choice: Oral or topical antihistamines are usually are effective. Complications require other interventions.
Reservoir Hosts: Many mammals such as cats, dogs, opossums, raccoons, squirrels and other rodents.
Control Measures: Insecticidal treatment of dwellings, pets, and/or of adjacent facilities frequented by flea-infested animals. Flea collars may help to control flea populations on pets. Repellents such as DEET help to reduce the number of flea bites.

Etiology/Natural History

Adult fleas are bilaterally flattened, small (1-8 mm) insects with sucking mouthparts for piercing the host integument to initiate blood feeding. Most of the world's approximately 2200 described species of fleas feed on mammals or birds but not on humans. Fleas undergo a complex (holometabolous) life cycle in which the immature stages differ morphologically from the adults. Eggs laid by gravid females usually drop from the host and lodge in the host nest or dwelling (especially in carpets) where the legless, eyeless larvae later hatch and develop. Most larvae feed on organic matter in these habitats using their powerful chewing mandibles. Some larvae, including those of the cat and dog fleas, consume protein-rich dry blood pellets that drop from the host after being excreted by the adults during blood-feeding. Completely fed larvae produce a flimsy silken coccoon to which debris often adheres before they molt to the pupal stage. "Pre-emergent" adult fleas may hatch from their pupa but remain inside the coccoon until a host stimulus (odor, vibrations, etc.) is detected. Pre-emergent fleas inside cocoons can survive for weeks or months remaining protected against desiccation but ready to break free rapidly when a host becomes available.

Epidemiology

Humans may be attacked by fleas under several circumstances but most of these involve the presence of companion or wild mammals in the immediate vicinity. If fleas are not controlled on companion animals, it is almost impossible to avoid flea bites.

Pathogenesis

Tissue and immunologic responses to simple flea bites are directed against salivary antigens (enzymes, etc.) inoculated by feeding fleas.

Clinical Manifestations and Complications

Most flea bites appear as uncomplicated, slightly itchy, reddened areas. A tiny purplish spot, or purpura pulicosa, at the bite site, surrounded by slightly swollen skin, or roseola pulicosa, characterize most flea bites. However, in some individuals, flea bites cause intense irritation for several days. Complications of flea bites include flea-bite dermatitis, an allergic reponse in which the bite from just one flea can cause dermatitis in sensitized individuals. Very rarely, anaphylaxis occurs. Other complications include the possible transmission of flea-borne pathogens such as the etiologic agents of plague (in western North America), murine typhus (also known as endemic or flea-borne typhus), cat-flea typhus and cat scratch disease. Cat and dog fleas are intermediate hosts of the double-pored tapeworm, *Dipylidium caninum,* which typically parasitizes cats and dogs but which can also parasitize humans.

Diagnostic Tests

Demonstration of biting fleas, or of fleas in the household or on companion animals. Diagnostic skin tests can identify flea bites in sensitized individuals including those with flea-bite dermatitis.

Treatment and Control Measures

A wide variety of insecticides (also called pulicides when used for flea control) can be used to combat flea infestations in dwellings and on companion animals. Access of reservoir hosts such as opossums and raccoons to pet food, garages, or crawl spaces under buildings should be prevented or limited. Flea-bite dermatitis can be treated by a series of desensitizing allergy shots. Flea-borne diseases require other clinical interventions.

Recent Advances and Contemporary Challenges

The advent of effective systemic and topical pulicides for cats and dogs has greatly reduced the abundance of fleas on these pets in areas where these drugs are widely used. Hopla (1982), Merchant (1990) and Genchi (1992) provide additional information on fleas that bite humans.

LOUSE FLIES (INSECTA: HIPPOBOSCIDAE)

Etiologic Agents: Several species including the sheep ked (*Melophagus ovinus*) (Figure 4), pigeon fly (*Pseudolynchia canariensis*), and deer keds (*Lipoptena* spp. and *Neolipoptena* spp.).
Clinical Manifestations: Bites, which may become reddened and swollen in some individuals.
Complications: Usually none; occasionally bites are painful or cause dermatitis
Mode of Transmission: Mainly from sheep, deer or pigeons, especially in individuals who work with these animals or live adjacent to herds or roosts.
Diagnosis: Louse fly bites are usually indistinc. Association of bites with louse flies is usually the most reliable diagnosis.
Drug(s) of Choice: Oral or topical antihistamine may be used, as needed.
Reservoir Hosts: Mainly sheep, deer and pigeons in North America.
Control Measures: Controlling these insects on their hosts with insecticides is usually the best control measure.

Etiology/Natural History

Most louse flies are dorso-ventrally flattened. They are a varied group including two families (Nycteribiidae and Streblidae) that feed exclusively on bats and are unlikely to be seen by most people. Louse flies are 2-12 mm long. The few North American species that bite humans, belong to the family Hippoboscidae. Adults may be fully- winged such as the pigeon fly and capable of flight, or wingless such as the sheep ked. Others, such as those associated with deer, initially have wings but these break off along suture lines after a host has been located. After engorging on blood, mated female flies deposit a single fully-developed larva near the host. The larva then quickly pupates and, weeks or months later, an adult fly emerges and searches for a host.

Epidemiology

Louse flies may bite individuals who farm sheep, keep pigeons, or frequent deer habitats. Sometimes, especially at dusk, deer keds will land on humans hiking in woodlands frequented by deer. The flattened shape and rapid, often sideways, running of these flies on the skin is distinctive.

Pathogenesis

Bites typically cause little pathology in humans.

Clinical Manifestations and Complications

Louse-fly bites are usually mild although dermatitis has been recorded in shepherds, especially during sheep-shearing.

Diagnostic Tests

None typically needed; association of louse flies with specific bites is diagnostic.

Treatment and Control Measures

Oral and topical antihistamines can alleviate bite itching, redness or swelling if this occurs. The best control measures are those undertaken to treat the reservoir hosts with insecticides. This often involves dipping of sheep. Repellents such as DEET deter louse-flies from biting humans and can be especially useful during sheep shearing.

Recent Advances and Contemporary Challenges

Since louse flies rarely bite humans, there are no major contemporary challenges.

MYIASIS-CAUSING FLY LARVAE (INSECTA: DIPTERA)

Etiologic Agents: Several species including the sheep nose bot (*Oestrus ovis*), horse stomach bots (*Gasterophilus* spp.), cattle grubs (*Hypoderma* spp.), screwworms (*Cochliomyia* spp.) (Figure 5A) rodent and rabbit bots (*Cuterebra* spp.) (Figure 5B), the human botfly or tórsalo (*Dermatobia hominis*) and the larvae of blowflies (*Calliphora* spp.).
Clinical Manifestations: Subdermal or intestinal infestation by fly larvae including the formation of a characteristic, often painful, "bot" or "warble" lesion in subdermal infestations that eventually has a small centralized opening through which the larva respires. Also, dead, gangrenous, ulcerated, or purulent flesh may be infested by blowfly larvae.
Complications: Damage to internal tissues and organs by early-stage migrating larvae; secondary infection in bot lesions may occur after the larva has exited.

Mode of Transmission: From adult botflies or blowflies ovipositing or larvipositing directly on human skin, from accidental contact with eggs near rodent or rabbit burrows (*Cuterebra* spp.), or from feeding mosquitoes carrying botfly eggs (*Dermatobia hominis*).
Diagnosis: Careful examination of lesions; this sometimes involves excision to remove and identify the bot.
Drug(s) of Choice: Usually none are needed because the bots secrete antibacterial chemicals into the lesion. Antibiotics may be required to prevent secondary infection after the bot has exited.
Reservoir Hosts: Mainly mammals, especially rodents, rabbits, sheep, cattle and horses.
Control Measures: Preventing flies, including mosquitoes, from landing on the skin usually by using topical repellents, window screens, bednets, etc. Also, avoidance of rodent and rabbit harborages is prudent. If handling rodents or their nest material, disposable gloves should be worn and fingers should not be touched near the mouth or nose until hands have been thoroughly washed.

Etiology/Natural History

Botflies, which include representatives of the dipteran superorders Muscoidea and Oestroidea, are a diverse assemblage of flies in which the larval stages are facultative or obligate parasites of vertebrates or feed on their dead tissue. Adult flies typically deposit eggs on the host, near it, or, in the case of *D. hominis*, on a mosquito carrier. These eggs hatch under a variety of circumstances but usually when they come in contact with host mucous membranes, or in the case of *D. hominis*, when its mosquito carrier, takes a bloodmeal. Females of the sheep nose bot propel larvae into the nose or eyes as they hover near the face. Female blowflies typically lay eggs directly on dead or purulent tissue and, despite the unsettling appearance of maggots feeding on human flesh, many of these are not true parasites and may be beneficial for wound cleansing. Early larval instars of most botflies (but not blowflies) undergo migrations through the host body before they arrive at the dermal site where they form a warble lesion. These sites are subdermal in most species, but in the alimentary canal in *Gasterophilus* spp., and in the nose or eyes in *O. ovis*. The bots feed on host tissue and fluids until they are fully fed (up to 5 cm in length), at which time, they exit the host and pupate in the soil before molting to adult flies.

Epidemiology

Infestation of humans by botfly or blowfly larvae is rare in North America. *Dermatobia hominis*, the screwworm (*Cochliomyia hominivorax*) and the secondary screwworm (*Cochliomyia macellaria*) were all once relatively common in parts of the United States but USDA control programs have now virtually eliminated them; they are still common in Central and south America however.

Pathogenesis

Despite their often gruesome appearance and relatively large size, most botfly lesions in humans are not pathogenic while the bot is present. Sometimes, a granulomatous mass forms at the warble site after the bot has exited.

Clinical Manifestations and Complications

A large, living, parasitic foreign object is present in the host. However, except for some fluid discharge, and pain (sometimes intense as the spiny larva moves around and feeds inside the lesion), pathology is often minimal. However, bots can, on rare occasions cause bleeding at the lesion site and secondary bacterial infections can occur after the bot has exited, especially in warmer climates or under unsanitary conditions. One rare but serious complication is the settling of bot larvae in atypical sites such as the brain before they molt to the next instar and start to grow; this can cause various neurological problems.

Diagnostic Tests

None usually needed for subdermal bots because these are visible externally. Bot antigen detection blood tests have been developed but are rarely used or needed.

Treatment and Control Measures

Because pain can be associated with bot infestations, it may be prudent to surgically remove some bots. Minor surgery is usually necessary to do this because most bots have backwardly-directed spines that lodge them in the lesion and prevent them from being easily pulled out. Other removal techniques that sometimes work include smearing vaseline or placing raw meat over the breathing hole; this may cause the larvae to slowly back out of the lesion as it tries to respire.

Recent Advances and Contemporary Challenges

The development of better avoidance methods would be helpful, especially against *D. hominis*, which can be a painful scourge in the adjacent neotropics. Hall and Smith (1993) and Goddard (1998) provide additional information on human myiasis.

SOFT TICKS (ARACHNIDA: ARGASIDAE)

Etiologic Agents: The spinose ear tick (*Otobius megnini*) and a few species belonging to the genera *Argas* (Figure 6) and *Ornithodoros*.

Clinical Manifestations: Reddened, often painful bites; larvae and nymphs of *O. megnini* may lodge in the ear.
Complications: Systemic reactions such as tick toxicosis, and transmission of the etiologic agents of tick-borne relapsing fevers; *O. megnini* may cause ear pain.
Mode of Transmission: Usually by sleeping near, or frequenting areas where rodent burrows, bat roosts, or chickens are common, or by working with livestock or chickens.
Diagnosis: Soft tick bites are often indistinctive but may be more painful and reddened than other arthropod bites; *O. megnini* can be located inside the ears.
Drug(s) of Choice: None usually needed unless tick toxicosis or relapsing fever occur, in which case, treating the symptoms or administering anti-spirochetal drugs, respectively, generally are effective.
Reservoir Hosts: Mainly rodents but also bats, livestock and poultry.
Control Measures: Control, relocation or elimination of rodent, bat, livestock and chicken populations adjacent to buildings; application of skin repellents such as DEET, when frequenting areas with abundant reservoir hosts is also advisable.

Etiology/Natural History

Morphologically, soft ticks measure from 1-15 mm in length, and can be distinguished from hard ticks because they lack a scutum (anterio-dorsal shield) and the mouthparts of nymphs and adults are situated in ventral grooves beneath the body, rather than protruding anteriorly as in hard ticks. The integument of soft ticks is also covered with small protuberances called mammillae whereas hard ticks have a relatively smooth external surface. Both soft and hard ticks have bloodfeeding larvae, nymphs and adults but soft ticks often have more than one nymphal stage. Soft ticks feed relatively rapidly (usually in minutes) and often while the host is sleeping or otherwise inactive so that many people who are bitten are unaware that ticks were present. Most species will feed on a variety of hosts including humans. The spinose ear tick is atypical in that only the larvae and nymphs feed and these usually lodge themselves in the ear canals of livestock or humans.

Epidemiology

Merten and Durden (2000) recorded 11 species of soft ticks biting humans in the United States (2 species of Argas, 7 of *Ornithodoros* and 2 of *Otobius*). Most bites occur in western North America where these ticks are more common. However one bat associated soft tick (*Ornithodoros kelleyi*) occurs throughout much of North America and will bite humans. *Ornithodoros turicata* occurs in Florida mainly in gopher tortoise burrows. In western states, people visiting rustic cabins or those camping outside may be bitten by soft ticks especially if rodent burrows are nearby. Because three western species of *Ornithodoros* can transmit relapsing fever spirochetes, indiviuals bitten by these ticks may become infected. The North American species of *Argas* that bite humans are normally associated with chickens.

Pathogenesis

Soft ticks typically inoculate a battery of chemicals when they feed which leads to localized tissue damage and sometimes to painful swollen bites. A salivary toxin can result in tick toxicosis or systemic complications.

Clinical Manifestations and Complications

Bites are often benign but can present as painful, reddened and swollen areas. This is common for spinose ear tick infestations. Blistering at the bite site, or systemic reactions can occur in tick toxicosis. A few human deaths have been recorded from soft tick toxicosis in North America.

Diagnostic Tests

Painful bites associated with ticks from sites with abundant rodents, bats, livestock or chickens indicate that soft ticks may be active. Being diagnosed with relapsing fever after outdoor activities in western states or provinces, also indicates that the individual was bitten by soft ticks.

Treatment and Control Measures

Uncomplicated bites can be treated with oral or topical antihistamine. Tick toxicosis requires careful monitoring and treatment of symptoms. Relapsing fever is usually treated with drugs such as tetracycline. Applying repellents such as DEET to the skin and avoiding, relocating or eliminating concentrations of rodents, bats, livestock and chickens also are advisable. Livestock can be treated for ticks using several methods including acaricidal dipping.

Recent Advances and Contemporary Challenges

Dusting of rodent burrows with acaricides in western North America for ectoparasite control shows promise in reducing bites by soft ticks. Hopla (1982) and Merten and Durden (2000) provide additional information on soft ticks biting humans.

HARD TICKS (ARACHNIDA: IXODIDAE)

Etiologic Agents: Several species but especially the lone-star tick *(Amblyomma americanum)* (Figure 7), American dog tick (*Dermacentor variabilis*) and blacklegged tick (*Ixodes scapularis*) in eastern North America, and the Rocky Mountain wood tick (*Dermacentor variabilis*),

Pacific Coast tick (*Dermacentor occidentalis*), and western blacklegged tick (*Ixodes pacificus*) in western North America.
Clinical Manifestations: Reddened, itchy areas around bite sites; these can persist for 2 to4 weeks after tick detachment by some species, e.g., the lone star tick.
Complications: Pathogen transmission that includes the etiologic agents of Colorado tick fever, Rocky mountain spotted fever, the ehrlichioses, Q fever, Lyme disease, tularemia, and human babesiosis. Secondary infection at the bite-site, systemic reactions, tick toxicosis, tick paralysis, and pathogen transmission.
Mode of Transmission: Most hard ticks quest for hosts from low vegetation. Hosts are usually vertebrates, especially mammals and birds.
Diagnosis: By detection of attached ticks on the skin, or by reddened bite sites after tick detachment.
Drug(s) of Choice: DEET on skin and permethrin on clothing are excellent tick repellents; permethrin also kills ticks if they contact it. Oral or topical antihistamine can alleviate bite reactions. Various drugs can be used to combat tick-borne pathogens.
Reservoir Hosts: Vertebrates, especially mammals and birds.
Control Measures: Ticks can be controlled in a variety of ways including habitat modification, treating domestic and wild animals with acaricides, and culling important hosts such as deer.

Etiology/Natural History

Hard ticks are common in most habitats in North America but especially in wooded areas. Ticks tend to desiccate in low humidity but some species are adapted to living in drier environments. The life cycle of hard ticks includes three blood-feeding stages, the larva, nymph and adult. Each stage feeds on a vertebrate host, engorges, detaches and then remains in the leaf litter for a period (ranging from days to months) before molting to the next stage and then seeking another host. During each blood meal, hard ticks stay attached to the host for an extended period, typically 3-7 days. North American hard ticks range in size from 1-15 mm.

Epidemiology

Merten and Durden (2000) recorded a total of 33 species of hard ticks biting humans in the United States; they also discussed records of additional species that may feed on people. Nevertheless, the main species that parasitize humans are those listed above and bites by one or more of these ticks can be very common in some parts of North America.

Pathogenesis

Hard ticks inoculate a variety of salivary chemicals when they feed; these disrupt host immune responses, form a blood pool and, in some species, a cement feeding plug.

Clinical Manifestations and Complications

Most tick bites itch in response to chemicals inoculated by the feeding tick. Some tick bites, such as those from the lone star tick, can remain reddened, itchy and swollen for several days. Complications include tick paralysis (mainly from the American dog tick), which is an ascending paralysis that resolves when the tick is removed. This paralysis is caused by a salivary toxin present in the saliva of some (not all) female ticks. Other complications include tick toxicosis and the transmission of pathogens such as those that cause Colorado tick fever, Rocky mountain spotted fever, ehrlichioses, Q fever, Lyme disease, tularemia, and human babesiosis. Most of these pathogens are transmitted by one or a few closely related species of ticks.

Diagnostic Tests

Presence, or removal of, attached tick. If the tick is retained in a small vial, it may be identified later if complications arise and may be useful in making a diagnosis.

Treatment and Control Measures

Attached ticks can be safely removed by grasping them close to the skin surface with fine tweezers and then gently pulling in a steady motion. Antihistamine may help to alleviate bite pruritis, swelling and redness. Tick paralysis is best treated by removal of the tick and careful monitoring of the patient, as needed, until full body movement returns. Tick-borne bacterial diseases are treated with antibiotics and supportive therapy. Topical repellents such as DEET and spraying clothing with permethrin also help to repel or kill questing ticks. Tucking light-colored pant legs into light-colored socks and having a tick body check after hiking are also prudent measures. Keeping to trails or mowed areas when hiking is also advisable. A variety of acaricides can be used to control ticks in the environment and on reservoir hosts. Seasonal burning also reduces tick populations.

Recent Advances and Contemporary Challenges

Many innovative techniques have recently been developed to help reduce tick populations. These include applying acaricides to wild deer and other animals at mechanical feeders, supplying acaricide-laced nest material for mice, and acaricide-laced bait stations for rodents and other small mammals.

Hopla (1982) and Merten and Durden (2000) provide additional information on ticks biting humans.

MITES (ARACHNIDA: ACARI), EXCLUDING CHIGGERS

Etiologic Agents: Several species of mites belonging to the arachnid subclass Acari, including the Chicken mite (*Dermanyssus gallinae*), northern fowl mite (*Ornithonyssus silviarum*), tropical rat mite (*Ornithonyssus bacoti*) (Figure 8), house mouse mite (*Allodermanyssus sanguineus*), straw itch mite (*Pyemotes ventricosus*), a bat mite (*Chiroptonyssus robustipes*) and three species of *Cheyletiella* (*C. parasitovorax* associated with rabbits, *C. yasguri* with dogs, and *C. blakei* with cats).
Clinical Manifestations: Small, reddened or itchy bite sites.
Complications: Allergic reactions including anaphylaxis; pathogen transmission.
Mode of Transmission: From infested reservoir hosts in or adjacent to human dwellings.
Diagnosis: Small bites in association with the collection and identification of the mites, or the presence of reservoir hosts.
Drug(s) of Choice: Orally or topically administered antihistamine.
Reservoir Hosts: Several birds and mammals, especially chickens, rats, mice, cats, dogs, rabbits and bats; also insect larvae (for *P. ventricosus*).
Control Measures: Acaricidal treatment of infested premises and relocation or elimination of reservoir hosts in the vicinity.

Etiology/Natural History

Most adult mites are relatively small (0.2-3mm) with immature stages being progressively smaller. Although there are a wide variety of life cycles in parasitic forms, most species have three active feeding stages, the larva, nymph and adult. In some species, larvae do not feed and in others, there is more than one nymphal stage. Most biting species feed on blood but some imbibe lymph or disrupted host cells. The straw itch mite normally feeds on insect larvae, especially those associated with straw, hay or grain, but it may also bite humans who handle these commodities. Although the scabies mite (*Sarcoptes scabiei*) has often been considered to be transmissable from animals to humans, recent research has shown that different mammal species are infested by different strains of this mite and that animal strains can only survive for short periods on humans.

Epidemiology

A wide variety of vertebrates are parasitized by mites and most species do not feed on humans. Those that do are associated with roosts or aggregations of their primary vertebrate hosts inside or close to human dwellings. If there is a host die-off or seasonal absence, or even if these mites are very common, they may bite humans.

Pathogenesis

Bite reactions result from host immune responses to antigenic salivary components inoculated by the feeding mites.

Clinical Manifestations and Complications

Bites may be numerous especially if the reservoir host population is large or has vacated the roost site. Dermatitis including hypersensitive and systemic reactions can occur. Bites from the straw itch mite cause a vesiculopapular dermatitis called "straw itch" that often results in large, reddened skin lesions. Also, *A. sanguineus* is the principal vector of the etiologic agent agent of rickettsialpox, and *O. bacoti* can harbor several human pathogens (Yunker, 1973).

Diagnostic Tests

None usually performed; the presence of mite-infested reservoir hosts and the association of mites with specific bites in humans can be diagnostic.

Treatment and Control Measures

Oral or topical antihistamine can alleviate bite reactions. Treating reservoir hosts or their roosts and bedding material with acaricides, and eliminating or relocating the reservoirs is also advisable. Straw itch can be difficult to avoid in persons whose occupations require handling hay, straw or grain. Application of topical repellents such as DEET can reduce the number of bites.

Recent Advances and Contemporary Challenges

The large array of mites that occasionally bite humans presents a daunting task with respect to precise diagnosis which can be important because some species transmit pathogens. Yunker (1973), Hopla (1982) and Merchant (1990) provide additional information on zoonotic mites that bite humans.

CHIGGERS (ARACHNIDA: TROMBICULIDAE)

Etiologic Agents: Larvae of several species (Figure 9), especially those belonging to the genus *Eutrombicula* (*E. alfreddugesi, E.batatas, E. belkini, E. cinnabaris,* and *E. splendens*); also, *Euschoengastia numerosa* in some western states.

Clinical Manifestations: Reddened bite sites that are extremely itchy in most individuals often for several days.

Complications: Extreme pruritus; allergic reactions including anaphylaxis; secondary infection at the bite sites which may heal slowly in some individuals.
Mode of Transmission: From vegetation, including grass, shrubs, leaf litter and fallen logs.
Diagnosis: Very pruritic reddened bite sites in most individuals; some individuals react less severely especially after many bites over several years. Bites are often clustered around belt lines or other clothing constrictions.
Drug(s) of Choice: Oral and topical antihistamine can help to alleviate the intense itching associated with bites.
Reservoir Hosts: Virtually all terrestrial vertebrates are reservoir hosts, especially for *Eutrombicula* spp.
Control Measures: Avoidance is often the best approach; repellents such as DEET and permethrin can be helpful.

Etiology/Natural History

Chiggers undergo three feeding stages during their life cycle, the larva, followed by the nymph and then the adult. However, only the tiny (< 0.5 mm) 6-legged larvae are parasitic. Eight-legged nymphs and adults are predators of small arthropods such as springtails or their eggs in the leaf litter and soil. Following mating, adult female chiggers lay clusters of tiny eggs in these habitats where the larvae later hatch and quest for a host. Host preferences of most chiggers are very broad and include humans. However, other chigger species are more host-specific and will not feed on humans. Some chiggers are more habitat than host-specific; for example, *E. splendens* is more abundant in moister environments. Feeding by larval chiggers involves host attachment followed by the formation of a specialized feeding tube called a "stylostome." The chigger then releases a mixture of enzymes and other chemicals into the bite site and later imbibes a combination of fluids and ruptured cells (not blood) from its host.

Epidemiology

Human-biting chiggers are almost ubiquitous in some woodland habitats especially in the southern United States. More than 50 bites are not unusual during a single visit to a chigger-infested area and sometimes hundreds of bites are recorded. Bites are concentrated in the warmer months when larval chiggers are most active.

Pathogenesis

The host recognizes the stylostome and inoculated chemicals as antigens which almost invariably results in intense redness and itching. Reservoir hosts generally experience much milder reactions at chigger attachment sites.

Clinical Manifestations and Complications

Multiple reddened, itchy bite sites are characteristic of chigger infestations. Because these bites may remain itchy for several days, scratching them sometimes results in secondary bacterial infections. No pathogens are known to be transmitted to humans by North American chiggers.

Diagnostic Tests

None typically needed because chigger bites are usually distinctive. Chiggers are so tiny that it is fruitless to search for them on the skin to implicate them as the cause of bites. Also, the chiggers have usually detached, been scratched off, or died, by the time a bite reaction becomes noticeable.

Treatment and Control Measures

Oral or topical antihistamine is often efficacious. However, for some people, nothing seems to alleviate the intense itching. Chiggers can usually be avoided by keeping to trails in fields and woods, refraining from sitting down in these habitats, using repellent such as DEET around the ankles, tucking pant legs into socks and spraying permethrin around these. Chiggers can be controlled in small areas using various acaricides.

Recent Advances and Contemporary Challenges

The development of reliable methods to alleviate the extreme pruritus that accompanies chigger bites in most individuals is a major challenge. Hopla (1982) provides additional information on chiggers biting humans.

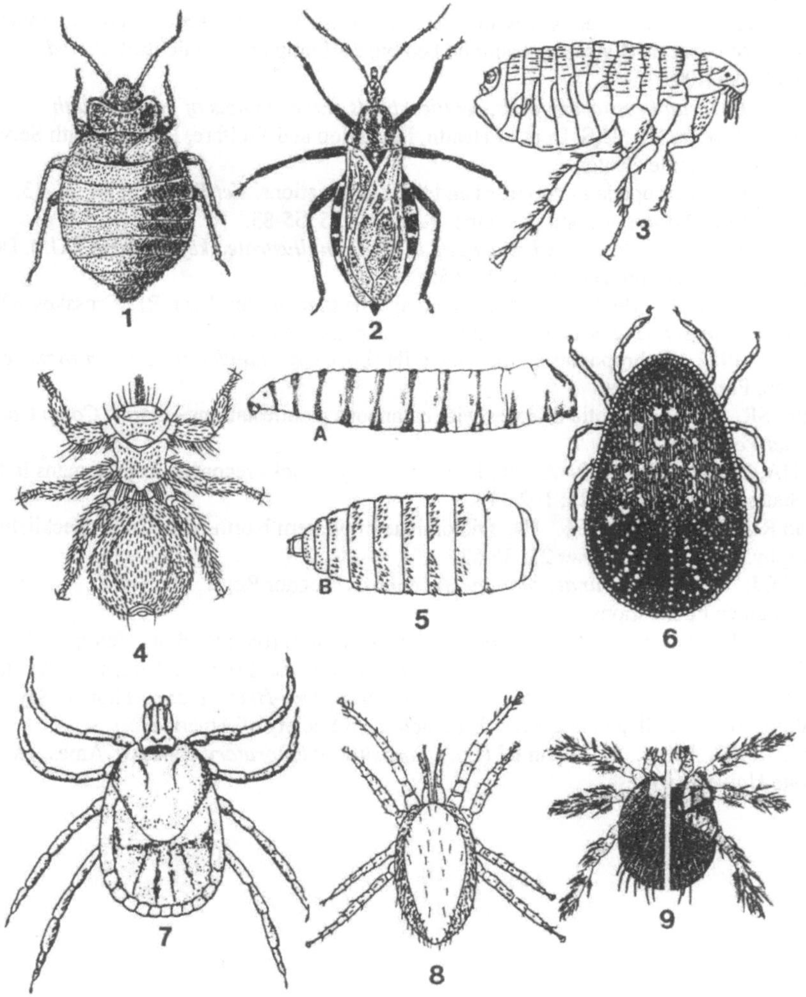

Figure 1. Bedbug (Cimex lectularius). Figure 2. Conenose bug (Triatoma sp.). Figure 3. Cat flea (Ctenocephalides felis), female. Figure 4. Louse fly (Melophagus ovinus). Figure 5. Myiasis-causing fly larvae; A-Secondary screwworm (Cochliomyia macellaria); B-Cuterebra sp. Figure 6. Adult soft tick (Argas sp.). Figure 7. Adult hard tick (Amblyomma sp.), female. Figure 8. Mesostigmatid mite. Figure 9. Chigger (Leptotrombidium sp.), larva (dorsal morphology on left, ventral morphology on right). Figures 1-4, 5B and 6-8 from CDC (1966); figures 5A and 9 from Gorham (1991). Drawings not to scale.

REFERENCES

Blow JA, Turell MJ, Silverman AL, Walker ED. 2001. Stercorarial shedding and transstadial transmission of hepatitis B virus by common bed bugs (Hemiptera: Cimicidae). *J Med Entomol* 38: 694-700.
CDC. 1966. *Pictorial keys: arthropods, reptiles, birds and mammals of public health significance.* Atlanta, GA: U.S. Dept of Health, Education and Welfare, Public Health Service, Communicable Disease Center.
Genchi C. 1992. Arthropoda as zoonoses and their implications. *Vet Parasitol* 44: 21-33.
Goddard J. 1998. Arthropods and medicine. *J Agromed* 5: 55-83.
Gorham JR (Ed.). 1991. *Insect and mite pests in food: an illustrated key. Volume 2.* U.S. Dept of Agriculture, Agriculture Handbook No. 655.
Hall MJR, Smith KGV. 1993. Diptera causing myiasis in man. In: Lane RP, Crosskey RW (Eds.), *Medical insects and arachnids*. London: Chapman and Hall.
Hopla, C. E. 1982. Arthropodiasis. In: Steele JH (Ed.), *CRC handbook series in zoonoses.* Boca Raton, FL: CRC Press.
Merchant, SR. 1990. Zoonotic diseases with cutaneous manifestations – Part I. *Comp Cont Educ Pract Vet* 12: 371-378.
Merten HA, Durden LA. 2000. A state-by-state survey of ticks recorded from humans in the United States. *J Vector Ecol* 25: 102-113.
Ryckman RL, Casdin MA. 1976. The Triatominae of western North America, a checklist and bibliography. *Calif Vector Views* 23: 35-52.
Schofield CJ. 1994. *Triatominae: biology and control.* Bognor Regis, England: Eurocommunica Publications.
Schofield CJ, Dolling WR. 1993. Bedbugs and kissing-bugs (bloodsucking Hemiptera). In: Lane RP, Crosskey RW (Eds.), *Medical insects and arachnids.* London: Chapman and Hall.
Usinger RL. 1966. *Monograph of the Cimicidae (Hemiptera-Heteroptera).* Thomas Say Foundation, Vol. 7, College Park, MD: Entomological Society of America.
Yunker C. 1973. Mites. In: Flynn RJ (Ed.), *Parasites of laboratory animals.* Ames, IA: Iowa State University Press.

INDEX

abdominal cramps 177
abdoiminal pain 29, 48, 79, 92, 101, 105, 119, 158, 180
acanthocephaliasis 103-106
acetominophen 50
adenitis 29
adenopathy 142
Aedes taeniorhynchus 93-95
AIDS 3, 7, 77, 113-128, 136-145, 158, 178-181
albendazole 31, 60, 64, 68, 92, 177, 180-181
alkaline phosphatases 159
allergic dermopathy 172
Allodermanyssus sanguineus 198-200
alopecia 65
Amblyomma americanum 196-198
amebiasis 10
American trypanosomiasis 168-172
amphotericin B 168
analgesics 50
anaphylaxis 63, 101, 186, 188, 190
Anaplasma 153
Ancylostoma braziliense 90-93
Ancylostoma caninum 90-93
anemia 28, 79, 139, 156, 158, 159, 171, 186
aneurysm 171
angioedema 101
Angiostrongylus cantonensis 5
anisakiasis 99-103
Anisakis simplex 99-103
anorexia 29, 79, 105, 157, 158, 172, 177
anticonvulsive drugs 60
antihistamines 186, 192, 196, 199, 201
antipyretics 50
appendicitis 101, 105
arachnoiditis 59, 60
Argas 194-196, 202
arthralgias 28, 157
arthritis 96
aspirin 50
asplenia 156, 158
asthma 12, 28, 103
atovaquone 144, 160, 161
azithromycin 122, 144, 145, 160, 161
Babesia bigemina 151
Babesia bovis 151
Babesia divergens 151-164
Babesia microti 151-164
babesiosis 141-164, 197
bacteremia 46
bacterial infection 63, 92, 193, 201
badgers 33, 93
balantidiasis 173-175
Balantidium coli 173-175
Bartonella henselae 7
bat mite 198-200
batbugs 185-187, 202
bats 186, 187, 195, 196
baylisascariasis 33-38
Baylisascaris columnaris 33
Baylisascaris melis 33
Baylisascaris procyonis 2, 16, 33-38
bear 93, 137
bedbugs 185-187
beetles 104, 106
benznidazole 172
bilirubin 159
birds 185-187, 189
  chimney swifts 186
  pigeons 191
  swallows 186
birdbugs 185-187
black flies 93, 94, 97
blastocystosis 13
blindness 12, 59
bloating 119, 177
blood transfusion 159, 161, 170
blowflies 192-194
bobcat 87, 98
Borrelia burgdorferi 153, 159
bot flies 192-194
bronchiolitis 180
bronchitis 28
Brugia beaveri 97-99
Brugia lepori 97-99
brugian filariasis 97-99
bruising 158
buffalo 62
Bulinus 89
Caenorhabditis elegans 51
Calliphora spp. 192-194
camels 62
Campylobacter jejuni 7
cancer 28, 59, 96, 99, 101, 117, 120, 140, 158
carbamazepine 60
carcinoma 67
cardiac arrhythmias 171
cardiac pseudotumor 29
cardiac transplantation 141
cat scratch fever 7, 190
cataract 29

Catostomus commersoni 86
cats 15-16, 23-33, 66, 75, 77, 85-87, 91, 130-136, 166, 190
cat flea 74, 77, 189-190, 202
cat mite 198-200
cattle 9, 62, 85, 137, 151, 152
cattle grubs 192-194
cattle fever 151
cellulites 98
cercarial dermatitis 88-90
ceviche 78, 100
Chagas' disease 3, 168-172, 188
chemotherapy 77
chest pain 63, 92, 96
Cheyletiella blakei 198-200
Cheyletiella parasitovorax 198-200
Cheyletiella yasguri 198-200
chicken mite 198-200
chickens 137, 186, 187, 195, 196
chiclero ulcer 166
chiggers 200-202
chills 157
Chiroptonyssus robustipes 198-200
chloroquine 161
cholera 2
chorioretinitis 138-140
Cimex lectularius 185-187, 202
Cimex pilosellus 185-187
Cimexopsis nyctalis 185-187
cimicid bugs 185-187
cirrhosis 63, 67
claim digger's itch 89
clarithromycin 144, 145
clindamycin 143-145, 159-161
coagulopathy 158
Cochliomyia hominivorax 5
Cochliomyia spp. 192-194, 202
cockroach 103, 104, 106
cod 100
cold sweats 96
coloboma 130
Colorado tick fever 197
coma 29, 158
conenose bugs 169-172, 187-189, 202
congenital toxoplasmosis 130, 139-143
congestive heart failure 158, 171
conjunctivitis 96, 103, 158
constipation 174
copepods 76, 86, 87
cough 28, 63, 92, 96, 100, 101, 157
covert toxocariasis 28-29
coyotes 66
creatine phosphokinase 50
creeping eruption 5, 90-93
Crohn's disease 101
cryotherapy 168
cryptosporidiosis 5, 7, 13, 113-128
Cryptosporidium canis 116, 124
Cryptosporidium felis 116, 124
Cryptosporidium meleagridis 116, 124
Cryptosporidium muris 116, 124
Cryptosporidium parvum 2, 3, 4, 5, 7, 8, 12, 16, 113-128
Ctenocephalides canis 74, 77, 189-190
Ctenocephalides felis 74, 77, 189-190, 202
Ctenodactylus gondi 130
cutaneous larva migrans 5, 90-93
cutaneous leishmaniasis 165-168
Cuterebra spp. 192-194, 202
cystic fibrosis 177
cysticercosis 12, 13, 57-60
cystitis 180
dapsone 141
deer 137, 153, 155, 162, 191
deer keds 191-192
DEET 187, 196, 198, 200, 201
dehydration 174, 180
depression 158
Dermacentor occidentalis 196-198
Dermacentor variabilis 196-198
Dermanyssus gallinae 198-200
dermatitis 91, 191
    bed bug-associated 186
    cercarial 88-90
    contact 90, 92, 103, 187
    flea-bite 190
    mite-associated 199
Dermatobia hominis 192-194
dermopathy 172
dexamethasone 60
diarrhea 5, 48, 49, 79, 101, 105, 114, 118-123, 173, 174, 176, 177, 178, 180
dientamoebiasis 13
diethylcarbamazine 31
diethyltoluamide 162
dimethyl phthalate 162
Diphyllobothrium latum 75, 76, 78-81
Diphyllobothrium pacificum 76
Dipylidium caninum 74-81, 190
Dirofilaria immitis 32; 93-97
Dirofilaria subdermata 93-97
Dirofilaria tenuis 93-97
Dirofilaria ursi 93-97
dirofilariasis 93-97
disseminated hymenolepiasis 74, 77, 79

disseminated strongylodiasis 3
dizziness 105
dogs 15-16,23-33, 34, 62, 63, 66-67, 75, 77, 85-87, 91, 94, 95, 165, 166, 169-172, 190
dog flea 74, 77, 189-190
dog mite 198-200
duck hunter's itch 91
dust mites 103
dysentery 173, 174
dyspnia 49
earthworms 24
echinococcosis
 alveolar 65-67
 cystic 61-65
Echinococcus granulosus 61-65
Echinococcus multilocularis 65-67
edema 47, 48, 50, 60, 188
ehrlichiosis 153, 197
encephalitis 36, 60, 138-141
Encephalitozoon 178-181
Encephalitozoon cuniculi 179, 180
Encephalitozoon hellum 179, 180
Encephalitozoon intestinalis 179, 180
encephalopathy 36
endophthalmitis 29
Entamoeba coli 4
Entamoeba histolytica 4, 5
enteritis 46, 100
enterobiasis 13
Enterocytozoon bieneusi 178-181
eosinophilia 28-31, 34, 36, 48, 50, 96, 101
eosinophilic meningoencephalitis 34, 37
ependymitis 59
epilepsy 12, 139
epiphora 139
erythemia 90, 96, 171, 186, 188, 190, 192, 197, 201
ethyl chloride spray 93
Euschoengastia numersoa 200-201
Eutrombicula spp. 200-201
expectoration 63
Fasciola hepatica 85
fascioliasis 85
fatigue 157
fecal-oral contact 62
fenbendazole 37
fever 28, 29, 48, 50, 63, 92, 96, 101, 119, 138, 142, 156, 157, 159, 171, 174, 177, 180
fish 76, 78, 86, 100, 102
Flagyl 174, 177
flatulence 79, 119, 177
flea-borne typhus 189-190
fleas 73, 74, 77, 78, 189-190
flies 191-194, 202
folinic acid 141, 144
food allergy 102
food poisoning 48
fox hounds 165
foxes 66, 67
frogs 87
fumagillin 181
furazolidone 177
Gasterophilus spp. 192-194
gastric ulcer 101
gastroallergic anisakiasis 101, 102
gastroenteritis 48, 100
geophagia 4, 26, 37
gerbils 66
Giardia intestinalis 175-178
Giardia lamblia 5, 6, 9, 16, 175-178
Giardiasis 5, 10, 13, 175-178
Gigantobilharzia 89
glioma 59
global warming 10
Gnathostoma 91
goats 137
gondi 130
grain beetles 73, 74, 77
Haematosiphon inodorus 185-187
haemoptysis 63, 96
hamsters 66, 77, 160
haptoglobin 156, 159
headache 29, 79, 157
hearing loss 139
heartworm 32, 93-97
hemiplagia 34
hemoglobinuria 159
hepatic capillariasis 29, 30
hepatic echinococcosis 63
hepatitis 141
hepatitis B virus 186, 188
hepatomegaly 28, 29, 63, 67, 171
hepatosplenomegaly 138, 171
hepatotoxicity 65
herring 100
Heterobilharzia 89
HIV virus 3, 7, 15, 16, 113-128, 136-145, 158, 178-181, 186, 188
hookworms 5, 29, 90-93
horse 62
horse stomach bot 192-194
house mouse mite 198-200
HTLV-1 3

human bot fly 192-194
hydrocephalus 59, 60, 130, 139
Hymenolepis diminuta 74, 76, 77, 80
Hymenolepis nana 74-77, 79-81
hyperesthesia 158
hypergammaglobulinemia 28
hypersensitivity reaction 28, 49-50, 100, 101, 103, 145, 168, 186, 188, 190, 199
hypertension
  intracranial 60
  portal 63
Hypoderma spp. 192-194
hypogammaglobulinemia 177
hypopyon 29
hypotension 156
immunocompromised persons 2, 7, 15-16, 74, 77, 79, 98, 113-128, 136-145, 158, 162, 176, 177, 178-181
immunodeficiency 2, 98, 158
immunosuppression 50, 59, 68, 74, 77, 121, 171, 177
intracranial calcification 139
iodoquinol 174
ischemia 156
ivermectin 37, 92, 106
Ixodes dammini 153
Ixodes pacificus 196-198
Ixodes ricinus 152
Ixodes scapularis 153, 154, 196-198
jaundice 67, 139, 158
keratoconjunctivitis 180
ketoconazole 168
kissing bugs 169-172, 187-189, 202
krill 100
lactate dehydrogenase 50, 159
Leishmania infantum 165-166
Leishmania mexicana 165-168
leishmaniasis 10, 165-168
lemmings 66
levamisole 37
lice 77
liver transaminases 159
liver transplantation 68
lipoma 99
Lipoptena spp. 191-192
louse flies 191-192, 202
lupus erythematosus 98
Lutzomyia anthophora 166
Lyme disease 153, 159, 162, 197
Lymnea 89
lymphadema 98
lymphadenitis 98, 138
lymphadenopathy 28, 98
lymphocytosis 171
lymphoma 99
mackerel 100
Macracanthorhynchus hirudinaceus 103-106
Macracanthorhynchus ingens 103-106
malabsorption 176, 177
malaise 119, 138, 142, 177
malaria 10, 154, 160
Maltese crosses 154, 160
mebendazole 31, 37, 50, 64, 68, 106
megacolon 171
megaesophagus 171
Melophagus ovinus 191-192, 202
meningoencephalitis 171
Mesocestoides 75, 78-81
Metorchis conjunctus 86
metronidazole 174, 177
Mexican chicken bug 185-187
mice 66, 74, 77, 135, 137, 152, 162
  white-footed 153, 155, 159, 168
Microbilharzia 89
microcephaly 139
microphthalmia 130
microsporidiosis 178-181
millipede 104
mites 198-202
moniliformiasis 103-106
Moniliformis moniliformis 103-106
mononucleosis-like syndrome 138
mosquitos 93, 94, 97, 99
murine typhus 190
muscular paresis 29
muskrats 66
myalgia 48, 119, 138, 157
myiasis 5, 192-194, 202
myocarditis 47, 49, 138, 140, 141, 171
myositis 29, 47, 92, 139
Nanophytes salmincola 85
Nassarius 89
nausea 63, 79, 101, 105, 119, 157, 158, 177, 180
Neolipoptena spp. 191-192
Neorickettsia helminthoeca 85
Neotoma albigula 166
Neotoma micropus 166
nephritis 180
Nerodia 87
neurocysticercosis 1, 13, 60
neuroretinitis 36
neuroses 186
neurotrichinellosis 49
neutropaenia 65

niclosamide 79, 81, 82, 106
nifurtimox 172
night sweats 138
nitazoxanide 122-123
northern fowl mite 198-200
northern rat flea 189-190
Nosopsyllus fasciatus 189-190
ocular larva migrans 12, 28-30, 34-37
ocular microsporidiosis 181
ocular toxoplasmosis 142, 143
ocular tumors 30
Odocoileus virginiana 153, 155
Oeciacus vicarious 185-187
Oestrus ovis 192-194
offal 62
Opisthorchis sinensus 86
opossums 75, 190
optic neuritis 29
orchiectomy 96
Orchopeas howardi 189-190
organ transplant 3, 140-142
Oriental rat flea 189-190
Ornithobilharzia 89
Ornithodoros 194-196
Ornithonyssus bacoti 198-200
Ornithonyssus silviarum 198-200
Otobius megnini 194-196
oxen 151
pallor 158
panniculitis 29
panthers 87
papilloma 99
parasitology education 10-12
paromomycin sulfate 81, 122, 177
pentamidine 161
Pentostam 167-168
Peromyscus leucopus 153
pericystectomy 64
peritonitis 105
permethrin 162, 187, 198, 201
petechial hemorrhage 48
petechiae 50, 158
pharyngitis 158
phenytoin 60
Phocanema decipiens 99-103
photophobia 139, 158
Physa 89
pica 4, 26, 35, 37
pigeon fly 191-192
pigs 9, 46, 50, 58-60, 62, 104, 137, 173, 175
pikas 66
piperazine citrate 37
plague 190
Planorbis 89
Plasmodium 160
pleuritis 63
pneumonia 28
pneumothorax 63
polar bear 49
pollack 100
polymyositis 138
polyneuropathy 172
porpoises 100
pulmonary edema 158
pulmonary infarct 96
praziquantel 60, 80, 81
prednisolone 50
prednisone 31, 60, 92, 98, 143, 144
pregnant women 7, 16, 139-143, 174, 177
primaquine 161
proteinuria 65
pruritis 90-93, 96, 186, 188, 190, 192, 197, 199, 201
Pseudolynchia canariensis 191-192
Pseudoterranova decipiens 99-103
pulmonary edema 156
Pyemotes ventricosus 198-200
pyrantel pamoate 37, 106
pyrimethamine 141, 143, 144, 161
pyrimethamine-sulfadoxine 161
Q fever 197
quinacrine 81, 161
quinine 159-161
rabbit bots 192-194
rabbits 34, 98
raccoons 9, 16, 33-35, 37-38, 75, 85, 87, 91, 93-95, 98, 104, 169-172, 190
respiratory distress syndrome 158
retardation
  mental 12, 139
  physical 105
rats 66, 74, 77, 104
red snapper 100
relapsing fever 195
renal failure 158
reticulocytosis 156, 159
retinal detachment 29
retinal hemorrhage 158
retinitis 30, 36
retinoblastoma 30
retinochorditis 143
rockfish 100
Rocky Mountain spotted fever 197
rodent bots 192-194

rodents 24, 66, 74, 77, 80, 130, 166, 168, 196
rickettsialpox 199
ringworm 7
roxithromycin 123, 145
salmon 78, 100
salmon poisoning 85
sandflies 166, 168
Sarcoptes scabiei 199
sashimi 78, 100
scabies mite 199
schistosomiasis 10
schistosomes 89
scotoma 139
screwworm 5, 192-194, 202
sea-bather's itch 89
sea lions 100
seafood allergy 102
seals 100
seizures 13, 29, 59, 60
sepsis-like syndrome 139, 141
septicemia 47
serpiginous lesion 92
sheep 24, 62, 85, 137, 191
sheep ked 191-192
sheep liver fluke 85
sheep nose bot 192-194
shrews 66
Simulium venustum 93
Sinusitis 180
skunks 33, 75
smallpox 2
smoked herring 100, 102
snails 89
snakes 87, 88
sodium stibogluconate 167-168
sore throat 138, 158
sparganosis 86-88
spiramycin 122, 142, 145
Spirometra mansonoides 86
splenomegaly 28, 158, 159, 171
squid 100
squirrel flea 189-190
Stagnicola 89
steroid therapy 3, 31, 50, 60, 98
strabismus 29, 139
straw itch mite 198-200
strongylodiasis 85
Strongyloides procyonis 91
Strongyloides stercoralis 3, 4, 6, 29, 85
Subcutaneous edema 171
subcutaneous nodules 87, 88, 95, 96
sulfadiazine 141, 143, 161
sulfonamides 144-145
sushi 78, 100
swallow bug 185-187
sweats 157
swimmer's itch 5
swimming pools 114
Taenia solium 6, 12, 13, 57-60
tapeworm infection 57-69, 73-82, 86-88
tenesmus 119
tetracycline 174, 196
thiabendazole 31, 37, 92, 106
thorny-headed worms 103-106
thrombocytopenia 139, 159
thrombosis 63
tick paralysis 197
tick toxicosis 195, 196, 197
ticks 151, 152, 157, 162
  black-legged 196-198
  American dog 196-198
  deer 153, 154, 159, 196-198
  hard 196-198, 202
  lone-star 196-198
  Pacific coast 196-198
  soft 194-196, 202
  spinose ear 194-196
tinidazole 177
tinnitus 105
TNP-470 181
Torsalo 192-194
toxemia 79, 105
Toxocara canis 4, 23-33
Toxocara cati 23-33, 130
toxocariasis 1,4, 12, 13, 23-33
Toxoplasma gondii 1, 4, 7, 8, 12, 15, 129-150
toxoplasmosis 1, 5, 12,13, 129-150
triamcinolone acetonide 31
Triatoma sanguisuga 187-189
triatomine bugs 169-172, 187-189, 202
Trichinella britovi 43, 51
Trichinella nativa 43, 49, 51
Trichinella nelsoni 43
Trichinella papuae 43
Trichinella pseudospiralis 43
Trichinella spiralis 8, 16, 29, 41-51
trichinellosis 6, 41-51
Trichobilharzia 89
trimethoprim-sulfadoxine 161
trimethoprim-sulfamethoxazole 141
trisulfapyrimidines 141
tropical rat mite 198-200
Trypanosoma cruzi 3, 168-172, 188
trypanosomiasis 168-172

tularemia 197
tumor 96
tuna 100
turkey 137
typhoid 2
typhus 190
ureteritis 180
urticaria 28, 90, 101
uveitis 29, 139
Vampirelopis nana 74
vasculitis 28
vasoocclusive symptoms 156
ventricular aneurysm 171
visceral larva migrans 4, 12, 28-30, 34-37
vitrectomy 31
vitreous abscess 29
voles 66
vomiting 48, 63, 100, 101, 105, 119, 158, 172, 174, 177, 180
walrus 49, 100
waterfowl 88
weight loss 79, 158, 177, 180
West Nile virus 2
whales 100
white sucker 86
woodrats 166, 168
xenodiagnosis 160
Xenopsylla cheopis 189-190
yellowtail 100

MIX
Papier aus verantwortungsvollen Quellen
Paper from responsible sources
FSC® C105338

If you have any concerns about our products,
you can contact us on
ProductSafety@springernature.com

In case Publisher is established outside the EU,
the EU authorized representative is:
Springer Nature Customer Service Center GmbH
Europaplatz 3, 69115 Heidelberg, Germany

Printed by Libri Plureos GmbH
in Hamburg, Germany